FORESTRY COMMISSION BULLETIN 97

Research for Practical Arboriculture

Edited by S.J. Hodge
Research Officer,
Department of the Environment Arboriculture Contract,
Forestry Commission

Proceedings of a Seminar held at the University of York,
2–4 April 1990. Forestry Commission, Arboricultural Association
and the Department of the Environment

LONDON: HMSO

© *Crown copyright 1991*
  First published 1991
  Applications for reproduction should be made to HMSO

ISBN 0 11 710297 0
ODC 270 : 946.2 : (410)

KEYWORDS: Arboriculture, Forestry

Enquiries relating to this publication
should be addressed to:
The Technical Publications Officer,
Forestry Commission, Forest Research Station,
Alice Holt Lodge, Wrecclesham,
Farnham, Surrey, GU10 4LH.

Front cover: A healthy lime tree improving the urban environment of Milton Keynes. *(39504)*
Inset: Steel rods being used to measure soil aeration around a plane tree in Milton Keynes. *(39411)*

Contents

			Page
Preface		S. J. Hodge	v
Summary			vi
Résumé			vi
Zussammenfassung			vii
Setting the scene			
1.	Opening address. Arboriculture: the role of a partnership	J. Chaplin	1
2.	Keynote paper. Research for practical arboriculture	D.P. Heathcoat-Amory MP	4
3.	Arboriculture: the research need	A.D. Bradshaw	10
Amenity tree establishment			
4.	Plant production: tree training	R.A. Bentley	23
5.	Use of water-retentive materials in planting pits for trees	S.J. Hodge	29
6.	Root growth, water stress and tree establishment	T.J. Walmsley, B. Hunt and A.D. Bradshaw	38
7.	Analysis of performance in semi-mature trees in relation to a high water-table	C.R. Norton	45
Trees in towns			
8.	Importance of soil physical conditions for urban tree growth	B. Hunt, T.J. Walmsley and A.D. Bradshaw	51
9.	A study of urban trees	S.M. Colderick and S.J. Hodge	63
10.	Benefits of amenity trees	J.R. Matthews	74
Tree stability			
11.	Recent storm damage to trees and woodlands in southern Britain	C.P. Quine	83
12.	Survey of parkland trees in southern England after the gale of October 1987	J.N. Gibbs	90

		Page
13. Tree stability	H.J. Bell, A.R. Dawson, C.J. Baker and C.J. Wright	94
14. Tree root survey by the Royal Botanic Gardens Kew and the Countryside Commission Task Force Trees	D.F. Cutler	102

Tree health

15. Air pollution and tree health in relation to arboriculture	J.E.G. Good	107
16. Acid rain: tree health surveys	J.L. Innes	120
17. Ash dieback in Great Britain: results of some recent research	S.K. Hull	129

Disorders of amenity trees

18. De-icing salt damage to trees and shrubs and its amelioration	M.C. Dobson	141
19. Watermark disease of willow	J.G. Turner, K. Guven, K.N. Patrick and J.L.M. Davis	152
20. Recognition and investigation of unexplained disorders of trees	R.G. Strouts	161
21. Recent advances in detection of wood decay	D.A. Seaby	168
22. Tree decay in relation to pruning practice and wound treatment: a progress report	D. Lonsdale	177
23. Stump fumigation as a control measure against honey fungus	B.J.W. Greig	188

Arboriculture safety

24. Safety of harness and sit-belts in tree surgery work	H. Crawford	199

Concluding remarks

25. A summary consideration	D.A. Burdekin	211
26. Final comments	J.C. Peters	214

List of delegates — 216

Preface

This Bulletin records the proceedings of a 3-day seminar organised jointly by the Arboricultural Association and the Forestry Commission and held at the University of York in April 1990. The need for such a forum for exchange of research results and practical information has become apparent with the success of two previous joint Arboricultural Association/Forestry Commission research seminars.

In 1980 the first seminar was held at Lancashire College of Agriculture and the papers were published as Forestry Commission Occasional Paper 10 *Research for practical arboriculture* (now unfortunately out of print). In 1985 the industry was updated on developments in research during another seminar held at York University, the proceedings of which were published as Forestry Commission Bulletin 65 *Advances in practical arboriculture*.

With continued commitment to, and funding of, arboricultural research, particularly by the Department of the Environment, the range of papers submitted for the 1990 seminar has been impressive, exploring many 'avenues' of arboricultural practice and providing valuable information for the refinement of these practices. An attendance of 290 at the seminar and publication of this Bulletin are positive steps towards ensuring that research results get to where they are really needed; to practitioners, Local Authorities and other organisations involved with amenity tree planting and management.

The views expressed in individual papers are those of the individual delegates and not necessarily those of the Forestry Commission, the Department of the Environment or the Arboricultural Association.

S. J. HODGE (Editor)
June 1990

Research for Practical Arboriculture

Summary

This Bulletin records the proceedings of a seminar held at York University in April 1990, organised jointly by the Arboricultural Association and the Forestry Commission. The seminar was the third of its kind, held every 5 years, since 1980, updating the arboriculture industry on current arboriculture research in the United Kingdom. Twenty-six papers are presented under the section headings: setting the scene, amenity tree establishment, trees in towns, tree stability, tree health, disorders of amenity trees, arboriculture safety, and concluding remarks.

Recherches pour l'Arboriculture Pratique

Résumé

Ce Bulletin presente les comptes-rendus d'une séance à l'université de York en Avril 1990, organisée conjointement par l'Arboricultural Association et la Forestry Commission. Cette séance quinquennial était la troisième depuis 1980, avec le but d-informer l'industrie sur les recherches courantes dans l'arboriculture en Le Royaume-Uni. Ici on reproduit 26 études sous les sections: décrivant la scène; l'établissement des arbres pur l'agrément; les arbres dans les villes; la stabilité des arbres; la santé des arbres; les maladies des arbres d'agrément; la sûreté dans l'arboriculture; et les remarques finales.

Forschung für Baumzuchtpraxis

Zusammenfassung

Dieses Bulletin berichtet die Verhandlungen eines Seminars in der Universität York in April 1990, das durch die Arboricultural Association und Forestry Commission zusammen veranstaltet wurde. Dieses alle fünf Jahre wiederkehrende Seminar war das dritte seit 1980 mit dem Zweck die Baumzuchtindustrie über den gegenwärtigen Stand der Baumzuchtforschung in Grossbritannien zu unterrichten. Die 26 Vorträge sind hier in den verschiedenen Abschnitten wiedergegeben, nämlich: den richtigen Rahmen geben; Begründung von Stadt- und Landschaftsbäumen; Bäume in Städten; Baumstabilität; Baumgesundheit; Krankheiten der Stadt- und Landschaftsbäumen; Sicherheit bei der Baumzucht; und abschliessend Bemerkungen.

Setting the scene

Paper 1

OPENING ADDRESS Arboriculture: the role of a partnership

J. Chaplin, Chairman of the Arboricultural Association

Minister, Ladies and Gentlemen, may I, on behalf of the Association, the Department of the Environment and the Forestry Commission, welcome you all to our joint third major Conference and Seminar devoted to research. The first at Preston in 1980 and the second in April 1985 here at York, established the precedent for this one and no doubt, we hope, for more in the future. You will know, Sir, the resultant effects of the two recent storms, of October 1987 and January 1990, and personally as well, because your former family home at Knightshayes Court in Devon has suffered serious loss. This will have reinforced for you, as it has confirmed for the arboricultural profession, how all prevailing the forces of nature can be and how precarious our understanding of nature is. We are the inheritors of great exotic collections and widespread plantings, combined with our own native trees. Surveys show much of it is an ageing population, and certainly the storms have seriously damaged trees and woodland across southern England. It falls to arborists and foresters to clear much of this away and to replant, in order to restore our landscapes and townscapes for the future.

The how, when and why we do it brings us full circle to the need for knowledge of research and the need to work safely with this knowledge. This is the prime reason for some 300 delegates attending this Conference.

In 1972 our Annual Conference and Seminar resolved by acclamation 'to identify and evaluate problems associated with the cultivation and management of trees as individuals, or in small groups, and to make recommendations on research and other needs to resolve the problems'.

This remains as true today as it was then, and one of those 'needs' mentioned there led Professor Last's report of 1974 to recommend the dissemination of that research information. All the money invested in research is of little use other than to the self esteem of the researcher, unless it is widely and repeatedly put before the professional user and the wider public. Government sponsorship of research is wasted if Government does not also see that it is widely publicised. We therefore conclude that the Arboricultural Advisory and Information Service, founded by the Department of the Environment, under the wing of the Forestry Commission, at Alice Holt Lodge, as a result of the recommendations of Professor Last's report in the 1970s, was one of the most beneficial Government decisions for trees in that decade.

This has proved decisive in my opinion in helping to rebuff so much mystery, and ignorance associated with the establishment and management of trees, coupled as it has been with the broad work of the Arboricultural Association and many others.

We believe now, with hindsight, that the development of the Arboricultural Advisory and Information Service is crippled by a lack of resources as I detailed personally with our President, Mike Lord, at the Silver Jubilee Conference. It needs expansion and further consolidation and we hope the Review currently in hand will lead to positive development and growth of the service.

We now know that this profession is growing at a very considerable rate per year, in response to what can be described as the Green issues, or what is otherwise expressed as public concern for trees everywhere. As I said, in my address at the Silver Jubilee Banquet last September at Warwick, 'the perception of the public is undergoing a rapid and dynamic change, and long may it continue, but public concern has outstripped perceived Governmental action'. That is something the Government must retrieve. The recent Gallup Poll confirmed that 80% of the nation believes the Countryside is in danger and are dissatisfied with the Government's efforts to protect it, but research has, I submit, a part to play in retrieving that perception.

Eagerly awaited by the Association is the Government's response to the Batho Report, to which we made radical proposals for reform. We are satisfied that the Tree Preservation Order and Conservation Area legislation is on the whole beneficial, has the general respect of the public and is wholly worthy of retention. If there were any consideration of rescinding the legislation, I believe there would be widespread disbelief in Local Government, the profession and amongst the public. We want to see legislation evolving into a positive management instrument, instead of one wholly biased towards the stilted hand of preservation. The commitment of this Association, a charity, to advise government in further consultation still remains and we believe working consultation with you on this subject would be very beneficial. We have recently organised a number of technical seminars, and over 600 professionals have met on this topic, and the message coming back from those is in accordance with the principal viewpoint I have outlined. This is informed research, which in effect is the Association keeping its hand on the pulse and progress of events.

The Arboricultural Association is composed of commercial and Local Government members of all descriptions, backgrounds, concerns and interests. We provide the focal point for a unified professional interest and practice. The Association unfortunately has to work with a multi-faceted Government interface. The central interest of arboriculture, at Government level is, in our opinion, to be correctly found within your Department, Sir, but we also have to cope with Government decisions that rise in and from the Forestry Commission, the Ministry of Agriculture, Fisheries and Food, the Countryside Commission, the Scottish Office and the Welsh Office. Whilst as professionals we can cope with this, it is my submission that, for Great Britain as a whole, it is a severe national handicap, bad in organisational terms, and results in the lack of a politically cohesive direction. I will give a few broad figures to illustrate the problem we face. The Forestry Commission's *Facts and figures* for 1988-89 lists 172 000 hectares of unproductive woodland. The Forestry Act of 1919, 71 years ago established the principle of a 'national forest estate', now coming into real production, of a total of 920 000 hectares. Grants and tax aid have helped to restore and plant private plantations totalling 300 000 hectares, and new grant aid proposals are now in place and new hopes are expressed for urban forests, farm woodland and grant aid schemes, etc. Task Force Trees struggles to restore the 1987 storm damage, but, despite all this work over a period of 70 years, our total tree cover is only 10% in comparison with the continental average, which is double that. It is no wonder that we have an import deficit of £6 billion a year for timber: 21 million trees died in the Dutch elm epidemic, 15 million fell in the 1987 storm, 6 million in January this year. This illustrates a major national problem, and it requires a major national investigative research initiative, which can only be instigated and led by Government. I hope to see this sooner than later, and it would be supported by arborists and this Association. We have to solve this problem, and I believe that several diverse Government departments with responsibilities for trees is but one confounding factor. We do not have enough trees of any kind, whether in amenity or in forestry.

Research results are available to us to plant and establish them, but piecemeal initiatives by diverse Government departments and by individuals will not see enough into the ground, and thus never solve our inadequacy in this matter at the present rate of planting. Wherever the nature of this problem lies, we as a nation have to understand it before we can really begin to solve it, and we should remember how times have changed, as it is now very socially acceptable to conserve and plant trees, and to be seen to be doing so.

In the meantime this Association will continue to support the work of the independent Arboricultural Safety Council. Despite widespread commercial pressure for the Association to control the drafting of these Codes, the Council of this Association decided to ensure that the Safety Council became established as an independent body whose Chairman is Trevor Preston, a distinguished past Chairman of this Association and practising Senior Local Government Arborist. The Association members contributed £50 000 worth of time and expertise to the preparation and drafting of the Codes during 1989 alone, to the benefit of a much larger work force than is contained within the Association's membership. Much of our membership works in Local Government, and the commercial sector comprises individuals and very small companies in the main, who cannot contribute cash support to the financing of the Safety Council. Council practice is generally to provide these guidance notes for free. A large amount of work needs to be undertaken to sort out the real safety issues and we acknowledge the help of the Health and Safety Executive for their guidance and support, and for highlighting safety issues leading to the production of documents, not seen before in arboriculture.

This Association whole heartedly accepts the main thrusts of the Arboriculture Review Group of 1988. That report was a major achievement because it properly identified the size of the profession, and the areas of potential research problems. One of its most important conclusions was of course that we do not produce an economic end product. Our work produces long-term environmental improvement, which is of course of national importance, as environmental issues are now of the utmost interest and the subject of widespread public demand.

Whilst, therefore, Minister, we believe in the conservation of our healthy existing tree cover in urban and amenity situations, we also believe in the establishment and replacement of trees with new plantings. Whilst we believe in the retention of trees by legislation, and by example, we also believe this country should benefit from widespread expanded new plantings, and whilst all these matters are of vital importance, the means to carry them out have to be in place. Central to that is the education of staff, and central to their needs are research and the continuous dissemination of its results.

I know this Conference will enable the researchers to communicate with the delegates, and I hope in the questions and answers that the delegates and practitioners will communicate with the researchers for their benefit.

A vital need for the Department of the Environment, the Forestry Commission and the Arboricultural Association is for us to explore closer co-operation, and this Conference is one way for us to do it. We all believe, I am sure, that the spreading of knowledge and understanding is one of the most beneficial things we can do together. The Association gives thanks therefore for the research that the Department finances, but we want to see the recommendations of the Arboricultural Review Group adopted, and implemented as quickly as possible.

We therefore welcome you, Minister, and await with interest your speech.

Paper 2

KEYNOTE PAPER Research for practical arboriculture

D.P. Heathcoat-Amory MP, *Parliamentary Under-Secretary of State for the Department of the Environment*

Mr Chairman, ladies and gentlemen, I value the opportunity to address you today. Trees are your business – at a time when there is rapidly mounting interest in trees and the environment. Everyone in this audience will be aware of the growing public concern over the loss of the tropical rain forests – the effects of man on the environment if you like. Much closer to home, we have been vividly reminded of the forces of nature, through the effects on our landscape of the violent storms of January and February 1990.

Trees are important in all our lives. We plant trees as a crop to provide a whole range of products, from the roof structures of our houses to our daily newspapers. They also provide shelter for hill farmers' stock and, as we have recently come to realise, play a vital part in the great chemical cycles that affect our weather and the air we breathe. Even more fundamentally, trees provide spiritual uplift, whether in the traditional and historical forests of Sherwood and the New Forest or growing in our urban parks and gardens.

Trees, therefore, are centre stage. Arboriculture, the practical business of growing and managing amenity trees, and defining the relationship between trees and man, is very high on the Government's agenda. There is a tremendous groundswell of interest in the environment, and there are probably more people in Britain today interested in trees and keen to plant and grow them than ever before.

Let me start with the international dimension. Daily we read reports of the widespread destruction of forests throughout the world. In the early 1980s an area the size of England was lost each year, mostly in the tropics. Recent reports suggest that the rate of loss may have doubled.

People are concerned, and that concern is well founded. All of us are increasingly aware of the central role which forests and global tree cover play in determining the world's weather and the pattern of existence on earth. Forests form an integral part of our life-support system: for instance, they have the capacity to lock up large quantities of carbon. Destroying the forests releases carbon dioxide, a 'greenhouse gas', and contributes to the process of global warming. The average concentration of carbon dioxide in the atmosphere is now 25% higher than before the Industrial Revolution. Much of this has been caused by the use of fossil fuels, but deforestation is a major factor.

There are other effects. Loss of forests reduces evaporation, and consequently rainfall, in the regions concerned. Deforestation also leads to the loss of countless species.

Clearance of large areas of forest for other uses, especially by burning, is the damaging process. Properly managed forests, where harvesting of timber is balanced by growth or regeneration, will not add to the carbon dioxide problem.

The aim must therefore be to help developing countries to manage their forests on a sustainable basis. To that end we are co-operating with them and others through the UK's aid programme.

In 1988, the Prime Minister announced that the Government would aim to devote more resources under this programme to promote the wise and sustainable use of forests. And in November 1989, speaking to the General Assembly of the United Nations, she said that the Government would aim to commit a further £100 million of bilateral aid to tropical forestry activities over the next 3 years.

A year ago, the Overseas Development Administration was financing some 80 forest projects with a total value of £45 million. Now, 165 projects are up and running, or in preparation, with a value of £150 million.

Let us not forget the vast natural forests of the northern latitudes in the Soviet Union and North America. All forestry should aim to be sustainable, whether in the developed world or in developing countries. Our efforts must focus on using appropriate management techniques which will allow commercial forestry to proceed without reducing the forest resource that we will pass on to succeeding generations.

To be credible and carry conviction in voicing these concerns, we must recognise our responsibilities at home to maintain and, if possible, increase, tree cover.

At the end of the First World War, when the Forestry Commission was set up to promote the neglected interests of forestry and trees, Britain had less than 5% woodland cover. Sustained effort over the years has increased this forest cover to 10%.

This figure, however, takes no account of the 90 million broadleaved trees occuring as isolated trees, in groups or in hedgerows scattered over the rural landscape, which contribute to the timber resource. Nor, in terms of total tree cover, does it take account of the 'urban forest'. This has been estimated to be in the order of 100 million trees. Whilst these trees are grown primarily for amenity, this does not and should not preclude the productive use of their wood when the trees are felled.

It is our policy to increase total tree cover. The Government has raised the target for UK forestry planting to 33 000 ha per year. In addition, the Farm Woodland Scheme has been introduced to encourage farmers to establish woodland on land producing cereals and other commodities which may be in surplus. This aims to plant a further 36 000 ha over 3 years.

In the past, there has been considerable objection to forestry planting as a result of some insensitive choice of sites and species. Both commercial forestry interests and the Forestry Commission have recognised this. The Woodland Grant Scheme requires that environmental issues are taken into account, ensuring a much more considered and sympathetic approach.

There are two initiatives in which I have taken a special interest. First off the drawing board is the Countryside Commission's proposal for a new Midlands Forest spanning an area of 40 000 ha, which is intended to have the character of the Hampshire New Forest with its unique blend of wooded areas, farmland and villages. The purpose is to establish a lowland forest as a resource for the nation into the next century and beyond. It would be a major recreational and tourism resource, enhance the landscape and wildlife interest of the area and, in due course, contribute to national timber supply – truly multipurpose forestry.

There has been immense public interest in the proposal. As an environment minister, I am used to receiving many letters from the public asking me to prevent this or that happening. It is therefore a welcome contrast and delight to be on the receiving end of so much public enthusiasm. Parliament has debated the proposal and local campaigns have pressed the case of each of the five candidate areas.

Consultation on the Countryside Commission's ideas closed at the end of February 1990 and I expect to receive their proposals in the summer. All of you will appreciate that this will take time and careful thought by the professionals in the field and by the potential participants. I should emphasise that participation will be wholly voluntary, and that we must win the confidence and active participation of farmers, landowners, foresters, amenity societies and others.

The second initiative is to establish wooded areas – 'Community Forests' – on the fringe of major cities to provide contact with nature and to offer new leisure opportunities, just as Epping Forest does for Londoners. Other benefits would flow from these forests: derelict land could be restored and low-grade agricultural land could be put to more productive use whilst stimulating employment opportunities and recreation. By enhancing the local environment, urban areas would be better places in which to live and work.

So far, three Community Forests have been authorised, East London, South Staffordshire and South Tyneside, and, at the end of April 1990, my colleague David Trippier will be launching the Tyneside Forest.

Turning to more cataclysmic events, most of you, I am sure, will be aware of the work of Task Force Trees. We have been pleased to support this special unit of the Countryside Commission set up after the Great Storm of 1987 to assist in restoring the damaged landscapes in London and 16 southern and eastern counties. Following the recent spate of storms, we have agreed to an extension of the Task Force to include the areas newly affected. Where appropriate, grant aid towards restoration is available to all private and public landowners, including owners of historic gardens eligible for support from English Heritage. Such is the measure of our concern that, since the storm of 1987, the Government has allocated more than £15 million for the period 1992–93. This will help restore the damage to amenity trees caused by the storm.

In addition, we have recently announced the provision of another £1 million as the first tranche of a programme of aid to the areas which suffered losses this year. This week will see the planting of the millionth tree funded by Task Force Trees.

These steps reflect the value of trees to the nation, and the sums concerned are in addition to grants for storm damage replanting which have been made available to woodland owners from the Forestry Commission.

Whilst the natural reaction to such catastrophes is to replant quickly, we are stressing the need to take stock and avoid hasty, insensitive clearance, which can do great harm to nature conservation interests. Planting may be appropriate, but in some situations natural regeneration might be preferable. There is no undue rush to get trees in the ground. Where trees are not a threat to people or property, consideration should be given to leaving some of them so that they can continue to provide habitats for wildlife, allowing nature to heal the scars.

Government is committed to the importance of trees for amenity, and through its aid programmes has a direct interest in ensuring that this considerable grant expenditure achieves its objectives. In other words, we need value for money from our tree planting programmes, whether the money is from Government, private individuals, local authorities or funds raised by the public. My Department funds core research and an information service to these ends.

We all want the trees that we plant to be maintained in the expectation of a long, safe life. But one problem is that trees are often planted in inhospitable sites, such as streets with their denatured and compacted soils, or derelict land with its added problems of lack of soil or toxic materials and low fertility. Getting trees to grow on such areas is not at all easy.

My Department has recognised the problems such sites pose and has funded research to address specifically ways which will improve the viability and survival of such trees. For example, one of our research projects at Alice Holt has shown just how damaging rough handling and exposure of the roots in transit from nursery to planting site can be. Without care, the result can be permanent damage so that what is being planted is in effect a dead tree!

Similarly, competition for moisture for newly planted trees has been demonstrated to be, perhaps, the crucial factor in their survival and grass has been shown to be the most aggressive competitor. Grass and other weeds must be controlled if trees are to grow and flourish. Our research has identified critical timing and methods for reducing weed competition that can result in dramatic improvement in survival and enhanced growth.

Research alone is not enough. It is essential that research findings are widely and quickly communicated in ways which will ensure they are put to practical effect. In the case of amenity trees, these messages must reach a wide range of practical people. The range of potential recipients is well illustrated by the many professions and organisations represented by this audience today.

Our Arboricultural Advisory and Information Service based at Alice Holt has been in operation since 1974. It performs the day-to-day task of dissemination, supported by publications and seminars. Publications range from complete handbooks, such as *Trees and weeds,* through Arboricultural Leaflets, created to give detailed guidance on specialist subjects, to Arboricultural Research Notes intended as brief summaries of the current status of research topics.

All research reports are publicly available and we encourage researchers to publish their findings in a number of ways, especially in forms that are readily usable by practitioners. An excellent example is the publication based on the work carried out by Liverpool University, under the leadership of Professor Bradshaw, entitled *Transforming our wasteland.*

In addition to these more conventional means of communication, we have established 'demonstration plots' as living examples of good and bad practice in weed control. These have proved effective and others have joined us to sponsor sites, chosen for their high impact, to produce a total of 22 throughout the country. There are two nearby: one at the Great Yorkshire Showground, Harrogate, and the other at Askham Bryan College.

But, as I am sure you will realise, much remains to be done. The current programme will extend work in the areas of plant quality and growth of urban trees, incorporating some new work on measuring 'vitality'. To continue studies on prevention and control of decay and disease, we have commissioned expert pathologists in the Forestry Commission to produce an illustrated handbook to help identify diseases and disorders of trees. This will be of particular value in managing trees. In addition, we have used the destruction of the storms of 1987 and 1990 to good effect — in this case to provide information on the response of tree roots and species to storm conditions. Together with the Countryside Commission, the Department of the Environment has reacted positively to requests for additional funds to record and assess what might be called 'windfall' data. The findings from these studies are the subjects of presentations later today.

Other topics currently, or soon to be, under study include the socio-economic value of trees, the responses of trees on development sites and the extent and causes of damage to non-woodland trees. More widely, there is important research in hand on monitoring overall condition of tree populations.

Monitoring of tree health started in earnest in 1984 and the results have indicated significant changes from year to year — no doubt reflecting variations in the weather. Among other things, it is important to recognise two things. First, tree condition is judged in relation to the state of an ideal tree. Any particular species has a natural range of occurrence. The condition of a particular specimen relative to the ideal will depend on its geographical location, soil conditions, exposure to the wind and many other factors which may or may not be related to human intervention. Secondly, monitoring of tree condition can never, by itself, prove the reasons for the state of health that is found.

When serious forest health problems were discovered on the continent, many felt that air pollution was the cause. There is no doubt that air pollution can, under certain circumstances, damage trees, but we still have no firm evidence that air pollution is damaging trees in the UK. There are certainly many other factors that damage trees and the overall picture is highly complex.

My Department sponsors a substantial programme of research into the effects of air pollution on natural systems. We do very much want to know whether air pollution is one of the factors affecting trees in the UK, and I believe that our scientific understanding is now reaching the point where we shall be able to draw conclusions. We have, therefore, recently set up an independent scientific review group with the specific remit to examine the possible role of air pollution and other man-made stresses as well as natural causes in relation to observed changes in tree condition. We are looking forward to its conclusions later this year.

I hope this will lead to the end of the sort of situation that occurred last year when Greenpeace published a booklet entitled *Margaret's favourite places*, which claimed that trees in places well known to the Prime Minister were damaged by air pollution. These claims were checked, and, in all but one case, it was concluded that the damage was a result of adverse site conditions or the age of the tree.

But the work commissioned by the Department of the Environment is, of course, just part of the total range of research on subjects of direct relevance to arboriculture. We recognise that, whilst both the forestry and arboricultural industries are concerned with growing trees, the objectives and the problems facing each are often very different.

We look to the industry to help us identify research priorities. I attach great importance to the setting up of the Arboricultural Advisory Board with broad terms of reference and a wide membership including a range of Government Departments, Research Councils, professional bodies and industry representatives. The Board now provides a mechanism for the industry to set its own priorities and guide the research programme to reflect their needs. One of the review groups set up by the Board is assessing the demand for technical information including that for management of amenity woodland. I look forward to seeing its report.

In parallel the industry has taken its own initiative. The Arboricultural Safety Council was set up in 1988 by the Arboricultural Association to help establish safe working practices tailored specifically to those engaged in arboricultural work. Since then, the industry has acted rapidly in drawing up a range of safety guides. Within little more than a year, the Council has prepared the first five guides. This is important, since at least 10 000 people are employed in your industry.

I should like to congratulate the Chairman of the Arboricultural Safety Council, Mr Trevor Preston, and all the members for what they have achieved in so short a time. I know that this has engaged much time and energy of many people in your industry. The high priority that you afford this matter and the excellent reputation for safety that you have, is a clear mark of the professionalism of the industry and is a model of the approach recommended by the Health and Safety Executive. In recognition of this, I therefore have great pleasure in announcing that my Department is to support the publication and dissemination of the first 10 guides to a maximum of £10 000.

I turn now to a completely different area of interest. At times, tree preservation orders have been criticised for emphasising protection for old trees at the expense of management. This was neither the objective nor should it have been the effect. Rather, it should draw attention to trees of amenity value and in so doing stress to the owners the importance of inspection, care and maintenance. I commend local authorities that use the system for the *management* of their treescape, acknowledging the mortality of trees, rather than using such orders to *prevent* necessary work or change.

I also want to say something about our review of tree preservation orders. I know this is of special interest to people here today. Representatives of the Association – and, of course, of many other bodies – met our Reviewing Officer last year to give a detailed account of your views and ideas for changes in the present arrangements. In addition, your members are frequently engaged in work affecting trees protected by such orders. Although I cannot tell you today the detailed outcome of the review – which we are still considering – I do want to emphasise that the Government has taken careful note of the widespread support for tree preservation orders expressed on many sides. We are aware of the importance attached to the system by a large number of people. We would certainly need convincing before agreeing to the hasty removal of a system which is now well established and has, in the past, achieved good results. But this is a complex area of legislation. There is scope for improvement, amendment and streamlining, and we want to make sure we get this right. A further complication is that not only urban trees are involved. Trees in rural areas, including woodlands, also come within the terms of the legislation. The review is also looking at hedgerow aspects, a contentious subject in its own right.

So I hope you will bear with us if we take a little longer before we go public on how we see the future of arrangements for tree preservation orders. As professionals, you know the vital importance of taking all relevant factors into account before reaching these very important decisions.

But it is local authorities who are in the front line. Legislation has to go hand in hand with good practice. Therefore, I am heartened that the Association has taken a major initiative in promoting good practice in this area by holding a series of outstandingly successful seminars on this subject for local authorities. These have proved an important forum for discussing many of the issues covered by the Department's review of the legislation. They will assist us in ensuring that any new legislation is relevant to today's circumstances both in town and countryside.

Up-dating the Department's guidance on these matters is crucial. The existing circular *Trees and forestry* is undergoing revision and will be republished in the new format of Planning Policy Guidance notes.

Public concern for the environment has never been greater. The challenge before you, as the professionals in this field, is to harness this public interest and support, and increase your efforts still further to raise standards of tree care and management and to work together to nurture the trees and landscapes which we all cherish.

The British have always had a particular attachment to trees; some of the finest landscapes, gardens, and arboreta in the world are to be found in our islands. We owe it to future generations to continue this fine tradition.

Paper 3
Arboriculture: the research need

A. D. Bradshaw, *Department of Environmental and Evolutionary Biology, University of Liverpool, Liverpool, L69 3BX, U.K.*

Summary

There is a serious need for research to support the practice of arboriculture. The resource is comparable to that in forestry, and its value perhaps much greater. Until recently, most of its research support has come as a by-product from forestry. Yet, because its trees are grown for amenity and not production, in many different sites and conditions, arboriculture has a whole range of problems of its own. These problems are leading to poor performance in both newly planted and older trees.

Because the resource does not make money, but only spends it, research has seemed to many people to be unnecessary. But research is needed to improve the quality of the product and to make savings in expenditure. The latter could pay for any expenditure on research many times over.

Introduction

No profession or industry can work effectively without research support. In 1975, as a result of a working party set up by the Arboricultural Association, a formal policy of support for arboricultural research was initiated. Its subsequent history has been described recently at the Silver Jubilee Conference of the Arboricultural Association (Biddle, 1989). To make further progress, a Review Group was set up in 1985 by the Forestry Research Co-ordination Committee to examine the present and likely future needs for research in arboriculture. The group, consisting of Dr P.G. Biddle, P.E. Spurway and D. Patch and Professor A.D. Bradshaw, was chosen to represent different aspects of arboriculture. It met on many occasions between 1986 and 1988, and received large amounts of evidence. Its detailed report has been published (Forestry Research Co-ordination Committee, 1988). The object of this paper is to highlight some of the most important conclusions of the report and to suggest the directions that arboricultural research should follow in the next 15 years. It must be a personal view, although it portrays the findings of the Group.

The terms of reference of the Review Group defined arboriculture as 'the culture of individuals or small groups of trees, rather than trees in woodlands or forests'. The Group took the view that this meant, in practice, all trees grown primarily for amenity, rather than for production.

It is important, at the outset, to realise the implications of this: that attitudes to trees grown for amenity are essentially different from those grown for production. Because no profit can be generated and there is no overriding reason why amenity trees should be felled at maturity (indeed there are good reasons for not felling them then), there are few of the normal commercial pressures for efficiency on those who commission their planting. This is exacerbated by the fact that the people responsible, such as local councillors, have many other, conflicting, responsibilities and that there is now a great shortage of revenue money for maintenance, although there are often adequate capital sums available for establishment.

The result is that, in practice, most amenity trees in this country, after the initial establishment period, are badly cared for and are effectively orphans. By contrast, in countries with a strong forestry tradition the situation is very different, for example in the town of Wiesbaden in West Germany (de la Chevallerie, 1986). However, standards at establishment in these countries also leave a lot to be desired.

Financial value of the industry

All this leads us to a consideration of the value we place on amenity trees. Is it possible that we do not really value them at all? This seems very unlikely. Because of their growth and vulnerability, they cause us considerable trouble. It would be easy to get rid of all amenity trees and, as a result, save ourselves this trouble, but there is no sign of this happening. Wherever there are new opportunities, trees are planted, sometimes at great expense in difficult situations. What is wrong is that we have never assessed the value of amenity trees properly.

There are probably about 100 million trees in urban areas in Britain, on publicly and on privately owned sites. Careful surveys have shown that there are 65 000 trees in the London Borough of Wandsworth and 970 000 trees in Edinburgh.

Even with the system proposed by Helliwell (1967), it is difficult to put a value on these trees. The average amenity tree costs £5-40 to purchase and plant; the cost of planting an urban tree may be as much as £200. If we assume a conservative average cost of £20, the total cost of replacement would be £2000 million. This calculation, however, takes no account of the fact that most of our trees are not newly planted, but are mature, magnificent specimens, which in courts of law may be valued at £2000 each.

Another approach to the value of amenity trees is in the number of people who tend them: about 10 000 full-time-equivalent workers, with a potential wages bill, at £10 000 per person, of £100 million. In comparison, there are about 24 000 workers in forestry.

Finally, the sales of hardy nursery stock have a wholesale value of £121 million a year, similar to the gross forestry output of about £100 million a year. There are no measures available of the expenditure on all the other aspects of planting and management.

It is evident that we set a considerable value on amenity trees, though this cannot be readily quantified, and are prepared to spend money on them, even though there is no financial gain. We would not die or suffer seriously without amenity trees. Their equivalents are museums and galleries and sports facilities.

Further work is needed if we are to arrive at a proper valuation of the resource and the money we spend on it. Nevertheless, because of the already established magnitude of the expenditure and the value of the resource, it can be argued that we should be prepared to spend money to support the resource by research, to achieve a better product and to save money.

There can never be valid arguments against trying to achieve higher standards. But present governments often argue that it is only justifiable to spend money if it will make money. This, of course, cannot apply to a spending operation. But money saved on the costs of one operation is the equivalent to money made, since it becomes money available for other purposes. A 10% saving in expenditure on hardy nursery stock alone would be equivalent to a gain of £12 million a year, 16 times current expenditure on research on all aspects of arboriculture.

The other common argument is that the markets which benefit should pay. In a fully commercial situation this may be justifiable. But it

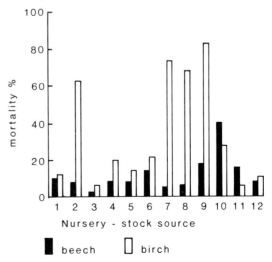

Figure 3.1 *Mortality of beech and birch stock purchased from 12 U.K. nurseries (from Kendle et al., 1988).*

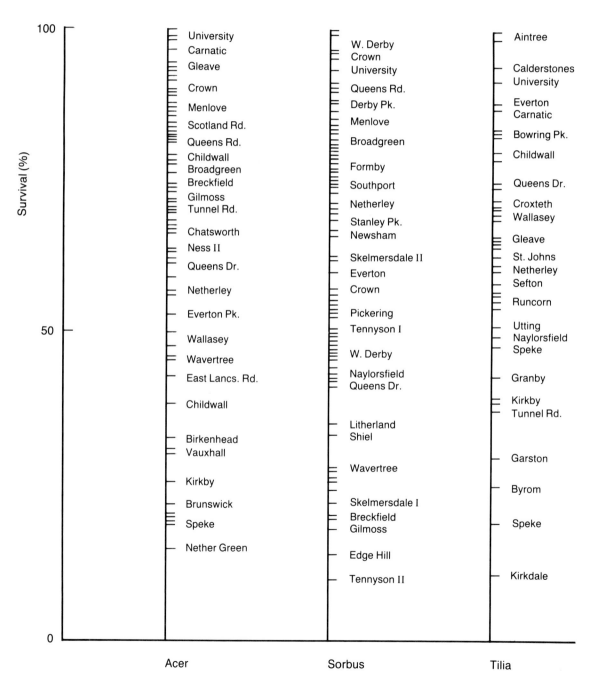

Figure 3.2 *Survival of trees 5 years after planting on 119 urban sites in the U.K. (from Capel, 1980).*

cannot apply either to private individuals or to local authorities. This is fully accepted for health research and other work for the common good. We expect Government to organise and pay for such work on our behalf out of the money we contribute in taxes.

The present situation

It is clear that all is not well with our approach to amenity trees. If car manufacturers or hospitals delivered products that performed to the same standard as amenity trees, there would be considerable concern. If manufacturers were involved, they would go out of business. The survival of new stock, for instance, when planted out with great care into excellent conditions, remains unpredictable (Figure 3.1); (Kendle et al., 1988).

The survival of trees after planting in urban areas is the acid test of the performance of the industry, since it sums the effects of many factors. Early evidence comparing what had been planted in a number of urban sites with what survived shows the magnitude of the problem (Figure 3.2; Capel, 1980). Recent continuous analysis of selected stands shows a process of cumulative attrition (Figure 3.3). This applies to replacements as well as to stock planted in different years (Figure 3.4) Gilbertson and Bradshaw, 1990). In Hampstead immediately after the Great Storm in 1987, it was possible to find almost as many young street trees standing dead because of bad practice as older trees fallen over because of the storm. Trees being strangled by ties are as easy to find in Westminster as in other cities. So, affluence does not get rid of bad technique. There is a serious lack of understanding of the problems involved in the management of amenity trees.

Despite this, the money being spent on research is abysmally small. The Review Group survey showed that in 1988 only £125 000 was being spent by universities and £615 000 by the Forestry Commission and the Department of the Environment, making £740 000 in total, compared with the £8 798 000 being spent on forestry research by the Forestry Commission. The amount has changed little since then. The

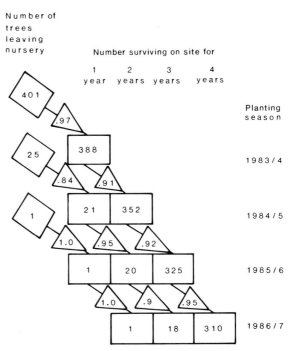

Figure 3.3 Life table for a total of 427 newly planted trees on six urban sites (from Gilbertson and Bradshaw, 1990).

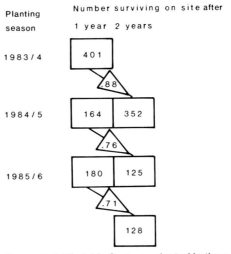

Figure 3.4 Life table for trees planted in three successive years on six urban sites (from Gilbertson, 1987).

annual expenditure on amenity tree research is only 0.6% of the sale value of hardy nursery stock. Most industries expect to spend at least 10% of their total turnover on research.

One great problem is the structure of the industry. First, there is a vast diversity of sites,

Table 3.1 Division of responsibility for planting and maintenance of most amenity trees in the U.K.

Operation		Responsibility
Overall plan		Commissioning body
Planning permission		Planner
Layout		Landscape architect (for commissioning body)
Planting		Contractor No. 1 (for landscape architect)
Aftercare:	first 2 years	Contractor No. 1
	subsequent	Contractor No. 2, direct labour, or **none***
		(for parks department or commissioning body)
	specialist	Contractor No. 3 (if invited)

* **N.B.** Most trees are orphans.

soils, and materials, but this is no different from agriculture. Secondly, there is fragmentation of responsibility: different organisations have responsibility not only for different groups of trees, but also for different stages of the life cycle of the same tree (Table 3.1). Within this pattern of changing responsibility, the people involved, in councils and committees, are liable to change every 3 years. This is very different from the patterns of management which have lead to the great landscapes from the past, which we admire so much.

Different Government organisations also have major involvement in amenity trees, including aspects of research, for example the Forestry Commission (FC), the Department of the Environment (DoE) including the Countryside Commission, the Natural Environment Research Council, the Ministry of Agriculture, Fisheries and Food, the Department of Transport, the Scottish Office and the Scottish Development Agency. This makes sensible planning and financing of research very difficult, although the Forestry Research Co-ordination Committee does what it can. It is to be hoped that the new Arboricultural Advisory Board will be able to produce a coherent programme and find adequate finance for it.

Research problems

Good detail on the outstanding problems requiring research is given in the report of the Working Group. Nevertheless, to obtain an overall picture of the research needs, it is necessary to examine the range of outstanding problems.

Policy and design

There are few clear ideas on the reasons why amenity trees should be planted, and there are few good analyses of their benefits, or their costs, in environmental terms. Yet their effects on pollution, noise, dust and wind are very positive. These are often quoted, but still need to be properly quantified. The sociological effects of trees are even less well defined, but must explain why we are so anxious to plant trees where we live and work.

This leads to a consideration of where amenity trees should be planted. Are streets the correct place, or should trees usually be kept to areas of greenspace? Much depends on their real benefits, as well as the problems they pose. There are no established guidelines and trees continue to be planted in unsuitable situations. Improvements will only be possible if detailed socio-economic analyses are undertaken – rather different from most arboricultural research. A start is the survey recently commissioned by DoE (Matthews, 1991).

At the moment, there is a major new initiative on community forests. It seems difficult for these to be planned properly without full research having been carried out on their potential benefits, and costs. However, little is available. Since it would be silly to delay such an initiative until the research has been done, the need now is to build research into the initiative, to provide a guide for the future. New developments such as this provide unique opportunities for critical research.

The need for, and absence of, proper management systems has already been mentioned. The trouble is that there is no analysis of the value of management systems for amenity trees. Recent analysis of the value of such systems for greenspace has shown that considerable benefits can accrue (Groundwork Trust, 1990). What is the value of inventories? In industry the realisation of their value has lead to transformations of industrial management systems. Similar benefits could be available to arboriculture.

Perhaps the most fundamental problem is the amount of money which should be spent on particular operations. This is a major question to be applied to all aspects of amenity tree work. It could, perhaps, be argued that the most economical management method is to plant the trees and then to leave them to grow by themselves, relying on their natural resilience, in the same manner as farmers treat large flocks of sheep. Present rates of tree survival do not provide any justification for this approach, and the results of death or poor growth are permanently visible on individual sites, casting a blight on them. This does not apply to sheep populations.

Selection of plant material

Nurseries list more than 1000 amenity tree species and cultivars, and catalogue descriptions abound. But a proper understanding of the performance of this material, equivalent to the detail known about forest species and provenances, is not available. There is little selection of material based on performance, because it is not known, and much is based on impressions, and even on fashion and ease of propagation.

There is a great need for a nation-wide survey of the performance of important species, backed up by performance trials in appropriate amenity situations. In agriculture and forestry the absence of such evidence would be unthinkable. In these industries, short- and long-term trials are routine and have been operating for decades.

Types of planting

It is curious that there is little or no hard evidence about the type of planting stock to select. Some authorities are adamant that only heavy standards can be used on their sites, although highly successful plantings based on whips have been established in the same region. What is disappointing is that little research has been carried out to establish which is best, particularly when the costs of such planting schemes can be so different (Michael, 1989) (Table 3.2), and the same paymaster is often involved, spending what is ultimately our money.

But the problem is not just whips v. standards. Other techniques, such as direct sowing, are being advocated. They may well be possible in some situations, but are only likely to be successful if the necessary research (e.g. Putwain

Table 3.2 Contrasting costs (at 1987/88 prices) of reclaiming and maintaining urban wasteland at three sites in Liverpool using heavy standard trees or transplants/whips. (Adapted from Michael, 1989).

	Cost ($£\ ha^{-1}$)	Total cost* ($£$)	Maintenance cost ($£\ ha^{-1}\ year^{-1}$)
Engineering approach Everton Park (24ha) grass and extra-heavy standard trees, on complete topsoil cover	188 500	456 700	11 100 (falling to £6800)
Ecological approach Crown Street (0.8 ha) grass/clover and transplant trees in mass, on brickwaste	15 500	12 400	900 (approx.)
Bamber Street (2.7 ha) grass/clover on brickwaste, transplant trees in mass on top soil	53 000	143 100	900 (approx.)

* Capital outlay discounted at 5%.

et al., 1988) is carried out. In the end, as with so much research for arboriculture, the need is to find the most cost-effective method, which implies that the research must be interdisciplinary.

Plant production

There are many aspects to plant production which need research so that growers can benefit by greater efficiency and better returns. But the overriding need, shown clearly in Figure 3.1, is for the growing and supply industry to deliver a product which works. The problem has been well recognised for 10 years (Insley, 1980). It is therefore disappointing that poor quality still occurs. What is needed is a reliable and simple test for quality. It is to be hoped that the current DoE/FC contract will be successful in this area of research.

There are other problems, such as those related to the use of container stock. There is little evidence that containers give better survival, and trouble may arise because of distorted root systems. In the end, what is needed are stock production systems that lead to high quality of performance for the user in the short and long term, and that are not oriented to benefit growers.

Establishment and growth

It has been apparent for some time that the conditions into which amenity trees are planted can be so different from those in which most forest trees grow that problems may arise. Indeed, it is remarkable that amenity trees in some situations, such as paved streets, perform as well as they do. The controlling factors, and therefore what must be ameliorated, are still not clear. There is evidence that both physical and chemical conditions can have overriding effects (e.g. Bradshaw, 1981; Jobling, 1981). But which is normally the most important is considered further at this conference (Colderick and Hodge, 1991; Hunt, *et al.*, 1991). To those who have advocated that one factor is overriding, it is clear that we must not think that we have all the answers.

Tree deaths soon after planting are a real problem. It has been argued that they could be

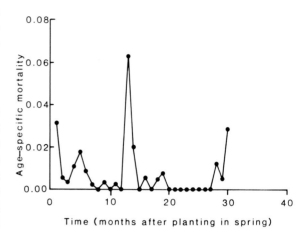

Figure 3.5 *Age-specific mortality of 401 newly planted urban trees (from Gilbertson and Bradshaw, 1990).*

related to water stress brought about by the often reduced root systems of newly planted trees. This is supported by a monthly analysis of mortality in newly planted trees (Figure 3.5). Deaths occur in the spring of each year. Real understanding can only come from detailed research, such as that by Walmsley *et al.*, (1991).

Our knowledge of competition and its effect has been transformed by the recent work of the Forestry Commission (Davies, 1987), which is an object lesson about the need for research in a topic where people thought they already knew the answers. It would have been a clever person who could have predicted the magnitude of the effects found. But, despite all the publicity, effective weed control is still rare in practice, suggesting that the need now is for research on how weed control can be provided most cost-effectively so that it can be automatic in all plantings.

The problem of support systems for large trees has not been solved. As long as some support is required, there is a need for a foolproof system.

An overall 'blueprint' system must be developed specifying the chemical, physical and other conditions required to ensure optimum establishment and growth. Such specifications have been determined for agricultural crops, but there is still a lot of work to be done before they are available for amenity trees.

Maturity and decline

Amenity trees are not only expected to grow in unnatural places, but are often retained to maturity or beyond. This raises many problems, with which arboriculturists are familiar. This has been highlighted by the recent storms. In commercial forestry, trees are felled at or before maturity and trees which are weak for any reason are removed as soon as they are noticed. With amenity trees, this is rarely acceptable. It is therefore necessary for inspection and management to have a proper scientific basis. One of the major problems is the recognition and treatment of decay. There is, in fact, a lot of research but the results are not yet readily available.

Damage and protection

Dutch elm disease (*Ophiostoma ulmi*) and now the outbreak of watermark disease (*Erwinia salicis*) in Dutch willows, (Turner *et al.*, 1991) warn us against complacency over disease outbreaks. We cannot merely pay attention to the diseases already present; we must watch for diseases which might appear and be prepared for them. Work on existing diseases, such as *Erwinia* (fireblight and watermark), *Armillaria* and other root-decay fungi must continue. For insect pests, the main need seems to be for constant vigilance.

Man, in his activities above and below ground, remains a major problem. There is a need for a revised version of British Standard 5837, controlling, rather than advising on activities on development sites. This needs to be supported by thorough research on the response of trees, and especially their roots, to damage. We know all too little of the abilities of trees to recover from damage.

Many agents can cause unnecessary damage. An example is the extensive damage caused by the indiscriminate use of strimmers. While there is a need for better management, we also need to prevent damage, by improving design. Although the incidence of damage due to vandalism is overrated (it is commonly the misinterpreted outcome of bad maintenance), the way in which design can reduce vandalism needs serious attention.

We have recently realised that serious damage is being caused by salt, but its extent has not been appreciated. The DoE desk study (Dobson, 1991) will be helpful in clarifying the situation and should point to further work by which both remedial and preventative measures can be developed. It is surprising, however, that this work is not being financed by the Department of Transport.

Effects of trees

Trees can damage services and houses, but it is not widely understood that most of this damage comes about because of faulty design or shoddy construction. Good advice is now available in NHBC Practice Note 3 of the National House-Building Council, but we need to examine further ways of preventing such damage by better design, and to ensure that this information is available to those who need it. The common view that trees should just be kept away from buildings and services is too simplistic and nihilist.

The beneficial effects of urban trees have already been mentioned. Until they have been quantified in relation to our physical and social environment, how can we plan the future?

Equipment and chemicals

Research on the materials and equipment used in arboriculture must mostly, unless questions of safety are involved, be the responsibility of the manufacturer, who stands to make a profit from them, but a system of validation is required.

Utilisation

The need for research on the utilisation of the products of arboriculture may not seem obvious. However, every year, a large amount of material is produced and removed. The cost of disposal of arisings may be half the cost of removal of a tree. Yet the material may be valuable in many different ways, and its disposal, if properly organised, could become a community industry. Most of the possible uses are already known, but have not been developed, and the situation is changing. For instance, the recent antipathy to the extensive use of peat in gardens suggests a market for wood chips as a substitute for peat as a garden mulch. The need is for managers to be educated as to what could be valuable in their situation.

Communication

Practical research is of no value if the results are not known and used. Communication must be seen as an integral part of research. The results must be published in a form that is available to, and understandable by, all those who could benefit by it. The comments of the Minister on this are welcome. Recently there have been some excellent examples of what can be done (e.g. Davies, 1987). Digests designed to make people aware of what is known are also important (e.g. Dutton and Bradshaw, 1982; Hibberd, 1989). Important results must be rapidly incorporated into relevant training programmes.

It is obvious, from the degree of poor performance in amenity trees, that much of what is already known is not being used, partly because of the breakup of the old apprentice system and the increasing employment, at all levels, of workers with inadequate arboricultural training. Whatever the explanation, lack of knowledge of established principles and recent research is the most serious problem connected with arboriculture.

Practitioners have, of course, a responsibility to be professional and to discover and pay attention to new findings. There needs to be a new attitude to practice. It is no longer satisfactory to do things the way they have always been done. Individuals must be aware of new approaches and techniques and be prepared to try them. Employers must realise that proper training, both at the outset and within career, are essential. There is an important, developing role here for the Arboricultural Association.

But there is inertia, and people are busy; the way must be eased. The initiation of the Arboricultural Advisory and Information Service (AAIS) as a result of the 1974 report, financed by the DoE, has been an invaluable step in the right direction. Both the advisory and the information aspects of the service have been very successful, but, unfortunately, it has remained substantially underresourced. A total of the equivalent of 1.5 technical staff supported by 1.5 administrative staff can just cope with day-to-day enquiries at the moment, despite a 20% growth in the number of enquiries received last year and the requirement to take on pathology problems. The service is much too small in relation to the size of the responsibility and the need to increase awareness of the proper handling and care of amenity trees. The AAIS is the major focus for improving standards, but it cannot do so without further development. More than 2000 people are employed in the service that supports agriculture.

Education and training programmes have a very important part to play in rapid and effective communication, but there is only one institution providing a full-time practical course in arboriculture (Merrist Wood Agricultural College); it is only part of courses in amenity horticulture elsewhere. This situation is not going to produce enough of the highly qualified specialists needed to work in local government and other organisations. Are the courses up to date? Are there enough courses providing good in-service training and are enough practitioners in arboriculture taking them? These are matters that the Arboricultural Association, as the professional organisation, should consider.

New techniques and good practice can reach a wide public through demonstrations. It is, therefore, good to see how quickly this suggestion has been taken up by the Forestry Commission and the AAIS with support from the DoE and local institutions in organising the 22 demonstrations around the country in colleges and show grounds.

Conclusions

The length of the list of problems highlights the need for further research. The resource we are dealing with is nearly the same size as forestry, but is potentially of greater value. Unfortunately, it has had very little research support, most of which has come as a by-product from forestry. The resource is undervalued because it does not make money. But it is expensive. Its economic and social value needs to be analysed more clearly.

Although there are problems in common with forestry, there are many problems which are special to arboriculture because the trees are being grown for an entirely different purpose and in very different situations. The performance of the product leaves a great deal to be desired, partly because of poor application of what is already known, but often because the problems are not clearly understood and the particular value of what is already known to be good practice has never been made clear by appropriate research. However, much of the poor performance is because the special problems that face trees grown for amenity have never been adequately analysed and appropriate techniques have not been devised to overcome them.

This third meeting on 'Research for Practical Arboriculture' shows the broad spread of problems and the progress being made in research to meet them. This situation is hearteningly different from the first small meeting held in Preston in 1980 (Forestry Commission, 1981), but we are still not doing enough. I hope it is apparent from this paper, and the full report of the Review Group on Arboriculture (Forestry Research Co-ordination Committee, 1988), that there is a great range of problems needing attention. Indeed, the situation is not likely to stand still; new problems will continue to arise.

Such research on an amenity resource cannot make money, but it can save a great deal and give better performance. If we want good, reliable products provided economically, we have to make sure that the necessary research is undertaken and that the results are properly communicated and put into practice.

ACKNOWLEDGEMENTS

This paper would not have been possible without all the time and wise counsel given so freely by my colleagues on the Review Group, Giles Biddle, Peter Spurway and Derek Patch, the evidence given to us by many people and the research of John Capel, Peter Gilbertson and Tony Kendle which has revealed much about the problems facing arboriculture.

REFERENCES

BIDDLE, P.G. (1989). Past, present and future of arboricultural research. In *Celebration of trees*, ed. J. Chaplin, 14–21. Arboricultural Association, Ampfield, Hampshire.

BRADSHAW, A.D. (1981). Growing trees in difficult environments. In *Research for practical arboriculture*, 93–108. Forestry Commission Occasional Paper 10, Forestry Commission, Edinburgh.

CAPEL, J.A. (1980). *The establishment and growth of trees in urban and industrial areas*. Ph.D. thesis, University of Liverpool (unpublished).

CHEVALLERIE, H. de la (1986). The ecology and preservation of street trees. In *Ecology and design in landscape*, eds A.D. Bradshaw, D.A. Goode and E. Thorp, 383–397, Blackwell Scientific Publications, Oxford.

COLDERICK, S.M. and HODGE, S.J. (1991). A study of urban trees. In *Research for practical arboriculture*, ed. S.J. Hodge, Forestry Commission Bulletin 97. HMSO, London.

DAVIES, R.J. (1987). *Trees and weeds – weed control for successful tree establishment*. Forestry Commission Handbook 2. HMSO, London.

DOBSON, M.C. (1991). De-icing salt damage to trees and shrubs and its amelioration. In *Research for practical arboriculture*, ed. S.J. Hodge, Forestry Commission Bulletin 97. HMSO, London.

DUTTON, R.A. and BRADSHAW, A.D. (1982). *Land reclamation in cities*. HMSO, London.

FORESTRY COMMISSION (1981). *Research for practical arboriculture*. Forestry Commission Occasional Paper 10. Forestry Commission, Edinburgh.

FORESTRY RESEARCH CO-ORDINATION COMMITTEE (1988). Arboricultural research: report of the Review Group on research on arboriculture. *Arboricultural Journal* **12**, 307–360.

GILBERTSON, P. (1987). *The survival of trees in inner city areas; a biological, social and economic analysis*. Ph.D. thesis, University of Liverpool (unpublished).

GILBERTSON, P. and BRADSHAW, A.D. (1990). The survival of newly planted trees in inner cities. *Arboricultural Journal* **14**, 287–309.

GROUNDWORK TRUST (1990). *Making the most of greenspace*. Groundwork Trust, St. Helens.

HELLIWELL, D.R. (1967). The amenity value of trees and woodlands. *Arboricultural Journal* **1**, 128–131.

HIBBERD, B.G. (ed.) (1989). *Urban forestry practice*. Forestry Commission Handbook 5. HMSO, London.

HUNT, B., WALMSLEY, T.J. and BRADSHAW, A.D. (1991). Importance of soil physical conditions for urban tree growth. In *Research for practical arboriculture*, ed. S.J. Hodge, Forestry Commission Bulletin 97. HMSO, London.

INSLEY, H. (1980). Wasting trees? – The effects of handling and postplanting maintenance on the survival and growth of amenity trees. *Arboricultural Journal* **4**, 65–73.

JOBLING, J. (1981). Reworked spoil and trees. In *Research for practical arboriculture*, 76–83. Forestry Commission Occasional Paper 10. Forestry Commission, Edinburgh.

KENDLE, A.D., GILBERTSON, P. and BRADSHAW, A.D. (1988). The influence stock source on transplant performance. *Arboricultural Journal* **12**, 257–272.

MATTHEWS, J.R. (1991). Benefits of amenity trees. In *Research for practical arboriculture*, ed. S.J. Hodge, Forestry Commission Bulletin 97. HMSO, London.

MICHAEL, N. (1989). Do you get what you pay for? *Horticultural Week* 12 May, 35–37

PUTWAIN, P.D., EVANS, B.E. and KERRY, S. (1988). The early establishment of amenity woodland on roadsides by direct seeding. *Aspects of Applied Biology* **16**, 63–72.

TURNER, J.G., GUVEN, K., PATRICK, K.N. and DAVIS, J.L.M. (1991). Watermark disease of willow. In *Research for practical arboriculture*, ed. S.J. Hodge, Forestry Commission Bulletin 97. HMSO, London.

WALMSLEY, T., HUNT, B. and BRADSHAW, A.D. (1991). Root growth, water stress and tree establishment. In *Research for practical arboriculture*, ed. S.J. Hodge, Forestry Commission Bulletin 97. HMSO, London.

Amenity tree establishment

Paper 4
Plant production: tree training

R.A. Bentley, *Pershore College of Horticulture, Pershore, Worcestershire, WR10 3JP, U.K.*

Summary

'Standard trees' produced in the U.K. are normally grown to meet the criteria stated in the specification for trees and shrubs in Part 1 of the British Standard 3936. Most deciduous tree species require detailed directive nursery training to form a product complying to the above standard. This produces an attractive tree with a well-developed crown, but arguably not ideal from the user viewpoint. A production method which encourages stem thickening at the expense of crown development is in widespread use on the Continent. Such 'light-crowned' trees may have advantages for tree producers and users.

Growth of nursery trees: the need for direction

This paper applies to the production of 'standard trees', but not to that of conifers, top-worked or container-produced trees.

Most tree species have evolved for life in a forest habitat, to be companions and competitors to other trees. Although in recent centuries man has made some selections, the growth habit of most trees remains similar to that of the wild ancestors. It is hardly surprising that the majority are not well adapted for life in the nursery and, if left unpruned, few would make handsome specimens.

The criteria specified for trees and shrubs in Part 1 of the British Standard (BS) 3936 (British Standards Institution, 1980) relate to stem circumference, overall height and length of clear stem. Most species also require a single central leader and a well-developed, evenly balanced head. In other words, the finished nursery tree should resemble a mature one in miniature, having a clear trunk and a developed crown. Such a tree seldom develops naturally.

Seedling trees in a mature woodland grow towards the light, endeavouring to reach and occupy an available space in the canopy. When they are 3 m tall, the appropriate maximum height for a BS 'standard tree', they are unlikely to have much of a crown and will probably be considerably thinner than the 8–10 cm BS circumference at 1 m from the ground. A sturdy stem is unnecessary for a tree that occupies the sheltered environment near the forest floor. For the best chances of reaching a space in the desirable, light-rich forest canopy, investment in height gain is the tree's favoured policy. With the exception of pioneer species, few wild trees enjoy the luxury of full light conditions in their infancy. Where man provides such conditions, a tree very different from the forest seedling is likely to result. The incentive to stretch for light has gone and thus, as long as grazing livestock are not present, a 'high-rise trunk' is a pointless investment. A giant shrub will be the result, the possession of single or multiple leaders depending upon the species. The main stem or stems are likely to be robust, and much larger than the BS requirements, but upward growth will be slow and the clear stem non-existent.

Wind swaying and a low crown will encourage stem diameter growth (Patch and Hodge, 1989). The overall effect will bear little resemblance to the BS specification for standard trees. The nursery environment, particularly for 1- and 2-year-old stock is similar to the open man-made habitat so, with reservations, prun-

23

ing to promote upward growth is the primary consideration.

Most of our ornamentals are worked, usually by budding on to a rootstock. The heading back of the rootstock at the start of the maiden year creates a root:shoot imbalance which leads to 'super vigorous' growth in the maiden year, which tends to negate the tendency for 'bush' formation in the first year. This is a major reason for the continued use of the budding technique, with all its inherent disadvantages, as a popular propagation method.

Pruning and training prior to crown formation

To prune a nursery tree directly and effectively, it must be growing well. High-quality trees cannot be produced on soils impoverished by poor nutrition, replant disease (Sewell, 1989), structural problems or impeded drainage. High incidences of pests, diseases and weeds must also be controlled.

A straight stem is important for budded or seed-grown trees and so upright planting is essential. Ridging up along the rows after planting helps on lighter soils. Following budding, bud guides to ensure a straight stem at the union can be applied at heading back (the aluminium type have given the most satisfactory results in trials at Experimental Horticulture Stations (EHS). Most trees, except *Aesculus*, *Fraxinus* and *Sorbus* spp. on sheltered sites, will need to be caned in their maiden year. Frequent tying of new maiden growth to the cane is required in the first summer. On exposed sites at Luddington EHS, nursery trees of *Sorbus aria* 'Lutescens' have proved more resilient to wind stem bending than *S. aucuparia* 'Sheerwater Seedling'. With the exceptions of the three genera listed above, most species also need their leading shoots cut back (commonly referred to as 'topping') at the end of the maiden year.

Topping checks apical vigour, thus diverting resources to crown and lateral development and giving an opportunity for increasing stem thickening. If the first-year growth is weak, topping can be delayed until the end of the second year.

An ideal topping height is 2.1 m for vigorously growing subjects such as *Acer platanoides*, *Laburnum* spp. and *Prunus* 'Kanzan', and 1.9 m for less vigorous subjects such as *Malus* spp., *Prunus subhirtella* and *Pyrus* spp. All *Tilia* species are weak growing in the nursery, and topping at 80–90 cm has been very successful for stem building in EHS trials.

The retention of a straight stem following topping at this height has not proved to be a problem for *Tilia*, but this could not be expected for all genera. It is advisable to cut off at least 30 cm when topping. If less is removed the regrowth will be weak and problems in maintaining a dominant leading shoot are likely to arise. Topping should be done just below a node, tying the new leader to a snag. The new leader is tied in to minimise a 'dogleg' kink in the stem. Timing of tying-in is crucial: the new shoots should be caught when they are about 3 cm long, when they can be bent in without a high risk of snapping off at the base. Early May is usually the best time and 1-inch masking tape is used to pull the shoot to the snag.

Failure to top trees such as *Malus* 'Profusion', *M. floribunda*, *Prunus* 'Kanzan', *Acer platanoides*, *Crataegus* spp. and *Ulmus glabra* can result in excessive apical growth at the expense of lateral development. Such trees become top heavy, thin in the stem and unstable. The masking tape and snag are normally removed in late summer. There are conflicting views on the merits of topping *Prunus avium* 'Plena', which seldom becomes unstable and thus achieving sufficient stem thickness is not usually a problem. However, if trees are left untopped, the leader tends to grow rapidly in the early summer. This rapid growth exerts a strong apical dominance, which effectively prevents the development of laterals. As growth slows down later in the summer the dominance is less strong and the upper laterals break; the season's leading growth could be described as a 'minicrown on a stick'. This, superimposed on the previous season's crown growth, results in the familiar and undesirable 'tiered' crown effect. Any check which slows the rapid early-season growth of the leader will negate or minimise the tiering effect. Topping or transplanting

Table 4.1 Effect of transplanting at the end of the maiden year on tiered crown formation in *Prunus avium* on Prunus 'Colt'. Rootstocks were planted in April 1985 and trees were assessed in December 1988. Data from MAFF, Luddington Experimental Horticulture Station.

Spacing between rows (cm)	% tiered	
	Not transplanted	Transplanted March 1987
40	24.7	2.4
50	23.5	9.5
60	32.4	8.3

Table 4.2 Effect of transplanting at the end of the maiden year on mean stem circumference at 1 m from the ground in *Prunus avium* on Prunus 'Colt'. Rootstocks were planted in April 1985 and trees were assessed in December 1988. Data from MAFF, Luddington Experimental Horticulture Station.

Spacing between rows (cm)	Mean stem circumference (cm)	
	Not transplanted	Transplanted March 1987
40	9.8	7.1
50	10.2	7.8
60	11.1	8.1

at the end of the maiden year (Table 4.1) will have a similar effect.

There is, then, a strong case for topping *Prunus avium* 'Plena' if no other action is taken to inhibit tiering. Although transplanting avoids the development of a two-tiered crown, trials with a range of genera have shown that it also results in reduced stem circumference in later years (Table 4.2). Root quality is seldom considered at the time of sale, this being much less easily quantified than stem circumference.

The treatment of temporary branches on the lower part of the trunk deserves special consideration. These help to thicken the trunk, so over-enthusiasm and too much removal, too, must be avoided. Some trees, such as *Prunus subhirtella* and many *Malus* spp., are prone to becoming bushy in the maiden year rather than making upright growth. Pinching all sideshoots back to 10 cm at regular intervals during the summer directs growth to the leader and produces a much better maiden tree. Trees grown from cuttings or by micropropagation have an even greater tendency to bushiness, so the sideshoot pinching technique must be applied. Any rootstock or basal stem growth should be removed as it arises.

At the end of the maiden year, any heavy branches should be removed, intermediate ones trimmed back to about two internode lengths and light ones left. Use a clean knife or secateurs and a suitable wound paint after winter pruning of *Prunus* or *Laburnum* spp., mainly to prevent the entry of silver leaf disease (*Chondrostereum purpureum*).

Forming the crown

At the end of the summer after topping the tree should be robust, with a framework of crown branches and some light lower-stem laterals. If the snag used to tie in the leader is still evident, it will require careful removal. At this stage the decision must be made on the crown type to be created. The U.K. market, guided by the British Standard, prefers a well-developed crown but the continental style of light-headed trees is gaining popularity. *Sorbus* species do not respond well to light-head pruning and the genus is unique in that the simple removal of low, unwanted branches at the end of each growing season creates a well-shaped tree. Too light a crown will not work well on the weaker growing Japanese cherries, so heavy pruning is to be avoided for these.

For a light-crowned tree, the crown branches are spur pruned back as near to the main stem as reasonably possible at the end of each growing season. Cuts are made just above a branch node. New laterals will then break from, or near to, the main stem. The exact position of the cut will depend on the species and the number of

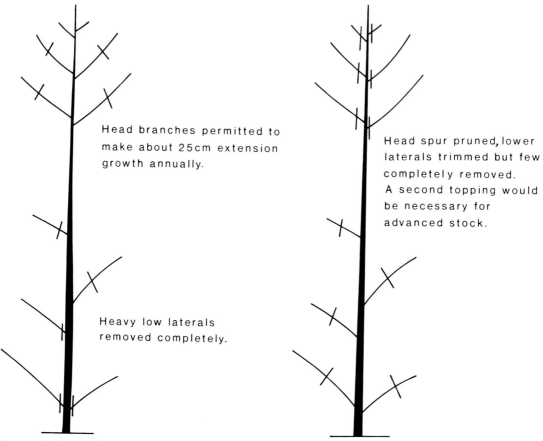

Figure 4.1 *Pruning for trees with a developed crown.*

Figure 4.2 *Pruning for trees with a light crown.*

previous cuts made to the lateral. The creation of 'blind' spurs which fail to regrow is not desirable. The method is repeated at the end of each nursery growing season, cutting to leave a tidy spur each time. It is important when applying this technique to leave several lower-trunk laterals at the end of each pruning. Failure to do this results in trees struggling to recover from such severe pruning and making insufficient stem growth each year. This method of pruning can be applied successfully to *Acer platanoides, Acer campestre, Carpinus, Crataegus, Laburnum, Platanus, Pyrus, Salix, Tilia* and most *Malus* spp. Species of *Aesculus* and *Fraxinus* will not regrow well from spurs or tipped branches, so they are best treated like *Sorbus* spp., but for *Fraxinus* spp. any lower branches would be retained for longer.

Before attempting any 'light-crown', continental-style pruning, U.K. nurserymen should check that such trees are acceptable to their markets. Such trees will have a crown of predominantly the current season's growth and will not, therefore, comply with the British Standard stating that 'The head shall be well developed for its type'. If such trees are not acceptable, about 25 cm of branch extension growth should be retained at the end of each growing season to produce a 'developed crown'. A little more growth may be acceptable in the lower part of the crown and less higher up. Cuts should be made just above a node and any potential competing leaders or crossing branches should be removed completely. The process is repeated each winter, allowing about 25 cm more extension growth each year until the desired size is reached. Most standard trees are saleable in the maiden plus two or three years. Heavy trunk

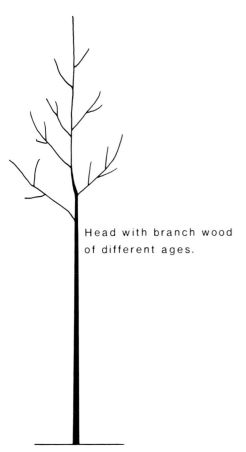

Figure 4.3 *Developed-crown tree ready for sale.*

Figure 4.4 *Light-crowned tree ready for sale.*

laterals are removed but lighter branches are shortened and left until the winter before lifting to encourage improved stem development.

In pruning for a light-head the lower lateral branches, some of which may be quite sturdy, are removed in the final June prior to intended sale. Providing the scars are not too large, this should allow sufficient time for callusing over before lifting.

The merits claimed for pruning for a developed crown (Figures 4.1 and 4.3) are that such trees

  i. look better immediately following planting,
 ii. have fewer unhealed scars caused by later removal of heavy lower lateral branches and
iii. are usually more acceptable to U.K. buyers and are thus more likely to sell or command a better price at present.

In light-crowned trees (Figures 4.2 and 4.4)

  i. smaller leaf area immediately following planting out results in low transpiration and improved establishment,
 ii. thinner lateral branches reduce the risk of vandals being able to lever the head over and break the main stem below the crown and
iii. light lateral branches can be tied to the main stem for easier, less bulky transport.

Conclusion

Standard trees produced in the U.K. are normally grown to meet the criteria stated in the Specification for trees and shrubs, British

Standard 3936, Part 1. Most deciduous tree species require detailed directive nursery training to form a product complying to the above standard. This produces an attractive tree with a well-developed crown, but arguably not ideal from the user viewpoint. A production method which encourages additional stem thickening at the expense of crown development is in widespread use on the Continent. Such light-crowned trees may have advantages for tree producers and users.

ACKNOWLEDGEMENTS

The information in this paper is based, in part, on development projects funded by the Ministry of Agriculture, Fisheries and Food.

REFERENCES

BRITISH STANDARDS INSTITUTION (1980). *British Standard 3936: Specification for nursery stock. Part 1. Trees and shrubs.* BSI, London.

PATCH, D. and HODGE, S.J. (1989). *Tree staking.* Arboriculture Research Note 40/89/ARB. DoE Arboricultural Advisory and Information Service, Forestry Commission.

SEWELL, G.W.F. (1989). *Replant disease and soil sickness problems.* Horticultural Development Council, Project News, No.7.

Discussion

T. La Del (Landscape Architect, Maidstone)

Budding appears to be undertaken for the benefit of nursery production not for the final users. Natural trees grown on their own roots may be slower to produce and establish but they could make a better product.

R.A. Bentley

Budded trees do make good maiden-year growth because the root:shoot ratio is unbalanced. Rootstocks have problems with incompatibility and sucker growth and are not necessarily ideal in the long term. The nursery industry is saying take it or leave it and not offering alternatives. Research is required on the long-term success of rootstocks.

Paper 5
Use of water-retentive materials in planting pits for trees

S.J. Hodge, *Forestry Commission, Forest Research Station, Alice Holt Lodge, Farnham, Surrey, GU10 4LH, U.K.*

Summary

Four types of water-retaining products, three based on polymers and one on seaweed, showed no benefits for survival or growth in five experiments on newly planted amenity trees. Intensive research on one of these products, Broadleaf P4 (a super-absorbent, cross-linked, polyacrylamide polymer), showed that it is only likely to be effective with a sustained regular water supply. This product may be of benefit in situations where irrigation can be assured and may indeed allow an extension of the interval between irrigation treatments, but, unless regular rainfall is experienced during the growing season, the product appears to yield no benefits in terms of survival or growth to newly planted trees that are dependent on rainfall alone.

Introduction

Manufactured products that claim to improve the survival and growth of newly planted trees by holding soil moisture against gravity and evaporation, whilst releasing it to trees, have been extensively advertised and are now being widely used. Agrigel, Growsoak 400 and Broadleaf P4 (which are super-absorbent, water-retaining polymers) and Alginure Soil Conditioner (a seaweed-based product with water-retentive properties) were tested in five experiments.

Three field experiments

Experiments (Expts 1 and 2) were established on a heavy clay site near Reading and a light sandy site near Thetford. No significant survival or growth response to the use of Agrigel or Broadleaf P4 was detected in any of the annual assessments for five growing seasons.

Because of the lack of response in these experiments, another (Expt 3) was set up on a light sandy soil near King's Lynn, to examine in more detail the performance of newly planted trees using Broadleaf P4 and Growsoak 400 incorporated at the manufacturers recommended rate into the backfill around oak (*Quercus*

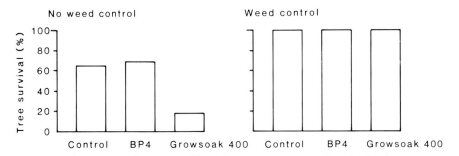

Figure 5.1 *Response of trees near King's Lynn two seasons after planting to Broadleaf P4 (BP4) and Growsoak 400 in the backfill at planting, without and with weed control. Survival means for* Quercus petraea *and* Alnus glutinosa; *(Expt 3).*

petraea) and alder (*Alnus glutinosa*) transplants. Weed control was superimposed on half of the trees to examine the relationship between these products, weed competition and tree growth.

After two growing seasons, no significant responses were detected to either product. However, as expected, intensity of weed control had a significant (99.9%) effect on tree survival and growth (Figure 5.1).

Two rain shelter experiments

Tests were carried out to determine the amount of water held by Broadleaf P4 in these experiments. First, an absorbency test was carried out using rain water and the water from the supply to be used to irrigate the experiments. Using rain water, Broadleaf P4 held 224 times its own weight in water against gravity; using irrigation water, it held 146 times its own weight in water against gravity. This difference in water-retention capacity is due to impurities, particularly calcium carbonate, in the irrigation water. Despite these differences, the results of both absorbency tests confirm the manufacturers' claim about the capacity of the unadulterated product to hold water against gravity.

A second test was undertaken to determine the available water-holding capacity, at various soil moisture tensions from field capacity to wilting point, of the soil in which the rain shelter trials took place and that of the same soil mixed with Broadleaf P4 at the manufacturers' recommended rate (Figure 5.2). The incorporation of Broadleaf P4 into the planting pit backfill made available 2.4 litres of water per 27-1itre planting pit in addition to the 2.4 litres of plant-available water that could be held by the soil alone, a doubling of the supply of plant-available moisture per tree at field capacity (excluding any supply of moisture from outside the planting pit).

To examine manufacturers' claims regarding water-retentive products more closely, two more experiments (Expts 4 and 5) were established using a mobile rain shelter to simulate drought conditions (Plate 5.1). A sensor on this equipment causes the shelter, which is mounted on runners, to move over the experiment as rain

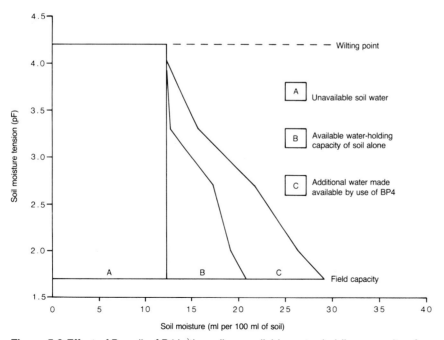

Figure 5.2 *Effect of Broadleaf P4 in the soil on available-water-holding capacity of the sandy soil, at Headley, Hampshire, on which rain shelter experiments were undertaken; (Expts 4 and 5).*

Plate 5.1 Rothamsted mobile rain shelter used in experiments to simulate drought conditions (Expts 4 and 5).

starts. After rainfall has ceased for 10 min, the shelter automatically retracts, uncovering the experiment. This allows precise control of the water available to trees in the experiment, but avoids the problems encountered in polyhouses with extremely high and variable temperatures and artificial humidities.

Gypsum blocks at 25 cm depth were used to monitor soil moisture tension in both experiments. The data collected are expressed as pF values; a standard method of expressing the tension with which moisture is held in the soil. At pF 0.05 (field capacity) the soil is holding as much moisture as it can; at pF 2.7, moisture is becoming less available to plants to the extent that plant growth begins to be limited by lack of water; at pF 4.2 (wilting point) no more water can be extracted from the soil by most plants. The higher the moisture tension, the less moisture is available to the plant.

From the second test, the manufacturers' claims appear to be valid in that Broadleaf P4 increased the plant-available moisture supply in the soil.

Experiment 4

The first rain shelter experiment examined the effect of Broadleaf P4 incorporated at the manufacturers' recommended rate into the planting pit backfill for lime (*Tilia platyphyllos*) whips. Trees were planted in $0.3 \times 0.3 \times 0.3$ m pits (27 litres) and 1 g of product per litre of soil was incorporated. The site was kept weed free throughout the experiment. All trees were watered to field capacity at planting and, subsequently, two rates of irrigation were applied: one keeping half the trees in soil near field capacity (irrigated) and the other keeping half the trees in soil near wilting point (unirrigated).

Despite Broadleaf P4 apparently increasing available soil moisture in laboratory tests, this was not reflected in the soil moisture tension and growth assessments made during the rain shelter experiments.

At the start of the unirrigated part of the experiment (Figure 5.3b), the soil in the control and Broadleaf P4 treatments was near field capacity. After 25 days, soil moisture tension under both treatments increased to above pF 2.7, the point below which poor moisture availability starts to limit tree growth. Beyond this point the unirrigated Broadleaf P4 treatment resulted in markedly higher soil moisture tensions (i.e. lower moisture availability) than the unirrigated control.

The reasons for this unexpected trend are not known. It would be expected from the laboratory trials that the use of Broadleaf P4 would

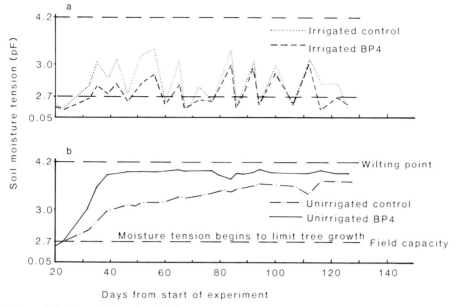

Figure 5.3 *Effect of Broadleaf P4 in the planting pit for* Tilia platyphyllos *on soil moisture tension, with and without irrigation; (Expt 4).*

maintain lower soil moisture tension (i.e. higher moisture availability) for longer than in untreated soil. The results of the second rain shelter experiment, however, (Expt 5) confirm these results.

The irrigated section of this experiment showed the opposite trend to the unirrigated section. Soil moisture tensions under the irrigated Broadleaf P4 treatment tended to be lower than under the irrigated control (Figure 5.3a). In practice, the technique used to maintain low moisture tensions in this experiment (irrigation with a fixed quantity of water irrespective of soil moisture tension after irrigation) failed to keep recorded soil moisture tensions below pF 2.7, the value which starts to limit tree growth. This has proved important in understanding how this soil ameliorant works.

Height and stem diameter increment in lime trees and total and mean shoot extension showed no significant response to the use of Broadleaf P4 (Figure 5.4). However, the results of growth assessments did tend to mirror the soil moisture tension data in that the unirrigated Broadleaf P4 treatment tended to result in lower tree growth rates and higher soil moisture tensions than the unirrigated control. Similarly, the irrigated Broadleaf P4 treatment tended to result in slightly better growth rates and lower soil moisture tensions than the irrigated control.

At the end of Expt 4, the root systems of all the trees were excavated. No obvious and consistent visual differences in root form and spread were noticed. All roots < 0.5 cm in diameter were oven dried and weighed, but there were no significant differences between root dry weights.

Experiment 5

The second rain shelter experiment examined the effect of Broadleaf P4 and Alginure Soil Conditioner incorporated at the manufacturers' recommended rate into the planting pit backfill of Norway maple (*Acer platanoides*) transplants. A basal fertiliser treatment was incorporated into the planting pit to isolate the water supply effect of these products from any fertilizing effect. Grass turf was laid over the site at the start of the experiment to intensify competition for moisture in what is a common situation in amenity planting schemes.

As in Expt 4, two moisture regimes were imposed on the trees: half received irrigation so that soil moisture tensions did not limit tree growth. The need to irrigate this part of the ex-

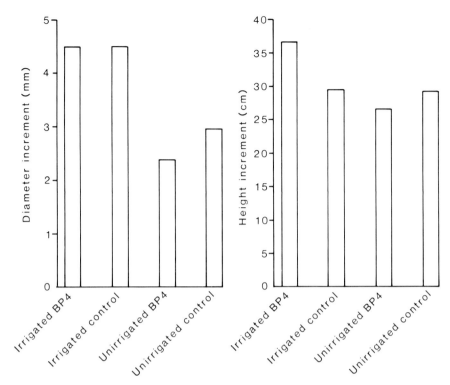

Figure 5.4 *Effect of Broadleaf P4 (BP4) in the planting pit on first year growth of* Tilia platyphyllos *with and without irrigation.*

periment was determined by gypsum block readings taken twice a week, and trees were trickle irrigated until all gypsum blocks showed the soil to be as near to field capacity as practically possible. The other half of the trees received an initial trickle irrigation to ensure that the ameliorated soil was as near field capacity as possible, but after that trees received no water at all.

As in Expt 4, gypsum blocks in the unirrigated Broadleaf P4 plots tended to show increased soil moisture tensions earlier than in the control plots (Figure 5.5b). Gypsum blocks in Alginure Soil Conditioner plots tended to show lower soil moisture tensions than in the other plots for up to 65 days.

At 70 days into this experiment, the automatic rain shelter developed a mechanical fault and rain fell on the unirrigated part of the experiment. Both Alginure and Broadleaf P4 plots held much more of that rainfall against gravity and evaporation than did the control plots. However, this high water-holding capacity did not noticeably extend the period over which this moisture was available to the trees.

As expected, there was a significant difference in height growth between irrigated and unirrigated trees, but there was no significant difference between amelioration treatments, although irrigated trees with Broadleaf P4 grew, on average, 1.6 cm more than the irrigated control.

Inevitably, leaves of trees in the unirrigated part of the experiment eventually turned totally brown (Plate 5.1). Rate of browning and time of total browning were recorded. Soil moisture tensions in the Broadleaf P4 and Alginure plots reached wilting point one week before the control. As a result, the onset of the total browning of foliage occurred a week earlier under these two treatments than under the control (Figure 5.5a).

At the end of the growing season, plants from each treatment were excavated to investigate the extent and characteristics of rooting, which were similar in all the amelioration treatments.

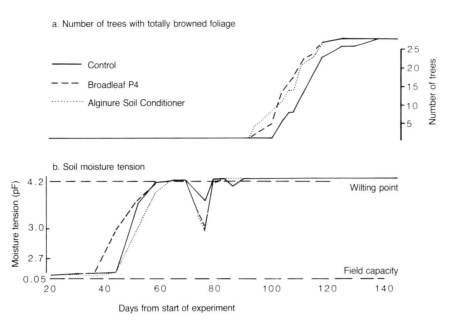

Figure 5.5 *Effect of Broadleaf P4 or Alginure Soil Conditioner on foliage browning of* Acer platanoides *and soil moisture tension, without irrigation; (Expt 5).*

Discussion

In addition to the work described, container trials have been carried out at Liverpool University (Woodhouse, 1989). These showed significant growth benefits to 2-year-old sycamore (*Acer pseudoplatanus*) and 1-year-old rooted cuttings of common osier (*Salix viminalis*) from the use of Broadleaf P4 incorporated into coarse sand at the manufacturers' recommended rate, with irrigation every 3 days or every 6 days.

No such benefit was detected in any of the five experiments described in this paper, despite Broadleaf P4 appearing to increase the available water-holding capacity of the soil in laboratory trials, probably because of differences in the frequency of water supply and moisture-holding capacity of the unamended substrate. The three experiments conducted under ambient conditions did not receive regular rainfall and in both of the rain shelter experiments, where irrigation was strictly controlled, the droughting treatment involved, in each case, only one low-volume irrigation in 130 days, whereas the Liverpool trials involved more frequent irrigation.

The growth rates of irrigated trees treated with Broadleaf P4, in both the rain shelter experiments, were better, though not significantly, than the control and the moisture availability under this irrigation regime was better for trees treated with Broadleaf P4 than for the control. The irrigation treatment did not succeed in keeping the very free-draining soil at field capacity (as can be seen from the data in Figure 5.3); the result was, in fact, a treatment similar to that administered in the Liverpool trials.

It appears that, for Broadleaf P4 to be of benefit to the survival and growth of newly planted trees, a regular input of water is required in order to 'recharge' the product. Any benefits of this product are most likely to accrue in the several days following recharging, when moisture is becoming less available in the soil. Continued hot dry weather, however, soon causes dehydration of Broadleaf P4 by transpiration. During a period of sustained drought, Broadleaf P4 soon becomes as dry as the surrounding soil and hence is of no benefit to the tree, and it requires a deluge to recharge the polymer.

It has been suggested that, on sites with poor weed control, moisture held in Broadleaf P4 might be utilised by aggressive grasses rather than by trees, thus possibly increasing competition for moisture and nutrients. This appeared to be the case in the King's Lynn experiment (Figure 5.1) where the competitive effects of weed competition were as severe in the Broadleaf P4 plots as in the control plots. Despite this obvious influence of weed competition, however, the lack of response from trees to Broadleaf P4 was consistent between experiments with and without weed control. This suggests that competition for soil moisture from vegetation is not the fundamental factor causing the lack of response to Broadleaf P4 in these experiments as compared with those at Liverpool University, although such competition clearly does improve the effectiveness of the product.

The results of the three field experiments under ambient conditions showed no benefit in terms of tree survival and growth from Agrigel and Growsoak 400 (cross-linked, polyacrylamide polyers). Any water-retentive properties of the seaweed-derived sodium alginate in Alginure Soil Conditioner had no beneficial effect on the survival of Norway maple transplants. Hummel and Johnson (1985) tested the water-retentive product Terrasorb (a gelatanised, starch-hydrolysed, polyacrylonitrile, graft copolymer) and found no significant benefit in its use on sweet gum (*Liquidambar styraciflua*) planted in a fine sandy soil. However, Walmsley *et al.* (1991) found significant growth benefits to sycamore (*Acer pseudoplatanus*) from the use of Aquastore (a cross-linked, polyacrylamide polymer) in a 1-year trial.

Conclusion

Despite much anecdotal evidence to the contrary (see Appendix 1), Forestry Commission research indicates that Broadleaf P4, Agrigel, Growsoak 400 and Alginure Soil Conditioner are only likely to be effective in improving moisture availability to newly planted trees in conjunction with a regular water supply. These products may be of benefit in situations where irrigation can be assured and may, indeed, allow an extension of the interval between irrigation treatments, but, unless regular rainfall is experienced during the growing season, they yield no benefits in terms of survival or growth to newly planted trees that are dependent on rainfall alone.

ACKNOWLEDGEMENTS
The work described was carried out with funding from the Department of the Environment.

REFERENCES
HUMMEL, R.L. and JOHNSON, C.R. (1985). Amended backfills: their cost and effect on transplant growth and survival. *Journal of Environmental Horticulture* **3**, 76–79.

WALMSLEY, T.J., HUNT, B. and BRADSHAW, A.D. (1991). Root growth, water stress and tree establishment. In *Research for practical arboriculture,* ed. S.J. Hodge. Forestry Commission Bulletin 97. HMSO, London.

WOODHOUSE, J.M. (1989). *Water storing polymers as aids to vegetation establishment in arid soils.* Ph.D. thesis, Liverpool University (unpublished).

Discussion

E. Guillot (Prospect Tree Services)
Did you find similar results with peat as you did with artificial ameliorants?

S.J. Hodge
Peat as with polymers generally showed no significant benefits apart from one site on a motorway embankment where growth improved when peat was used. One other site gave detrimental results, possibly because micro-organisms causing decomposition of the peat exert a strong nitrogen demand.

A. Mouzer (Rotherham Borough Council)
Last summer (1989) was very dry and watering was essential; if polymer ameliorants do not work, how much water should be applied and how often?

S.J. Hodge
Site preparation to produce a suitable soil structure is very important and thorough weed control is vital. The watering regime necessary will differ from site to site and

prescriptions are unrealistic. Trickle irrigation is preferable for maximum infiltration. Care should be taken to avoid rainfall run-off from roads and paving into planting pits because of possible high concentrations of pollutants.

A.D. Bradshaw (Liverpool University)

Autumn planting is good because the extent of rooting shows up to a fourfold increase compared with spring planting. Just as artificial ameliorants differ in composition, so does peat! Did you use peat with a high moss content or other types?

S.J. Hodge

Both moss and sedge peat were used.

E. Freeman (Rochester-upon-Medway City Council)

There are very chalky soils in my area. As Alginure has a high pH, can it cause a potential problem of very high alkalinity.

S.J. Hodge

No research was carried out on the effects of the high pH properties of Alginure; there is also the complication of high quantities of other elements in this product. My research was restricted to water-retention properties.

Appendix 5.1

In a postal survey undertaken by the manufacturers of Broadleaf P4, of the 92 respondents, 79% were able to compare P4-treated and untreated crops or plantings.

	Better or much better (%)	*No difference (%)*
Survival	89	11
Establishment	94	6
Growth rate	80	20
Growth quality	75	25
Root condition	72	28
Savings on watering	88	12

Paper 6
Root growth, water stress and tree establishment

T.J. Walmsley, B. Hunt and A.D. Bradshaw, Department of Environmental and Evolutionary Biology, University of Liverpool, Liverpool, L69 3BX, U.K.

Summary

Serious problems are likely to arise to newly planted trees because of severe restrictions to their water supply. The size of the root system of newly transplanted trees affects the volume of soil water they can exploit. Water stress can be minimised following transplanting, by ensuring that trees with large root systems are transplanted and encouraging the promotion of subsequent root growth. An alternative is the use of certain soil ameliorants by which the available water in the root zone can be increased. Both methods are important in reducing the number of times irrigation is needed.

Introduction

The root system of a tree is far more than an anchoring mechanism, it is the absorbing surface for water and nutrients, but it is this organ that is disrupted during transplanting. The amount of root system lost when the tree is lifted from the nursery is often in the region of 50% for young transplants (Kendle, 1988); for larger trees the value may be as high as 95% (Watson, 1987). It is therefore not surprising that the consequences of such root loss are severe. The tree will be predisposed to internal water deficits, as the root system will not be able to supply the transpiration demands of the relatively large shoot system (Witherspoon and Lumis, 1986). Stress may also occur from air gaps forming at the root-soil interface on planting, which will cause a resistance in the flow of water from the soil to the root, thus further reducing the amount of water available to the tree (Sands, 1984).

Although the nutrient status of a tree is closely linked to the water status, field evidence suggests that it is lack of water that is the principal factor leading to the stresses that eventually cause the death of the tree (Gilbertson and Bradshaw, 1990). This paper considers water stress after transplanting and the methods by which it can be reduced. In the light of a simple model of water requirements, two approaches were examined: the first is an increase in the volume of soil that the newly transplanted tree can exploit, by planting trees with larger root systems; the second is an increase in water availability, by irrigation or an increase in the water-holding capacity of the soil. With both approaches the main aim is to promote root growth and therefore to restore as quickly as possible a more favourable balance between the shoot and root systems.

Root growth and drought

Before examining the methods of reducing water stress and of promoting root growth of newly planted trees, it is perhaps pertinent to examine the effects of drought on root growth. The effects of water stress on shoot growth are well known as they are easily observable (as leaf wilt, or leaf abscission under severe conditions), but the effects on root growth are far less apparent.

Many investigations detail the gross effects of drought on root growth by reporting final root dry weights at the end of an experiment (Table 6.1). These investigations do not demonstrate the sensitivity of root growth to water stress. To accomplish this, continuous monitoring of root growth is necessary and then it can be shown

Table 6.1 Effects of drought on the growth of *Acer platanoides* transplants, 10 weeks from the start of a drought treatment.

	Total shoot Extension (cm)	Root dry weight (g)	Leaf area (cm²)
Droughted	87	19.56	1387
Watered	377	52.80	6145

that root growth is highly sensitive to water stress (Figure 6.1). Very soon after the onset of drought, root growth is disrupted.

If a tree suffers water stress on transplanting, it is in a vicious circle: it is unable to develop the root system with which it would be able to absorb the water which would relieve the stress. Field evidence suggests that this effect may well carry over from one year to the next, so that a tree which suffers from the effects of drought in the first year after planting may well be much more susceptible to drought in the second year.

A model of water requirement

The problems facing a newly planted tree are made clear by the simple model of Gilbertson *et al.* (1987), (Figure 6.2): the amount of water

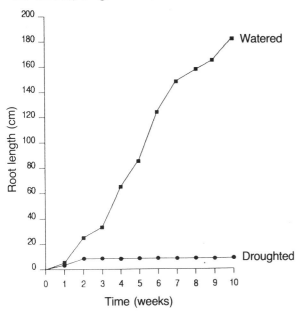

Figure 6.1 *Effects of drought on root growth of* Acer platanoides *transplants.*

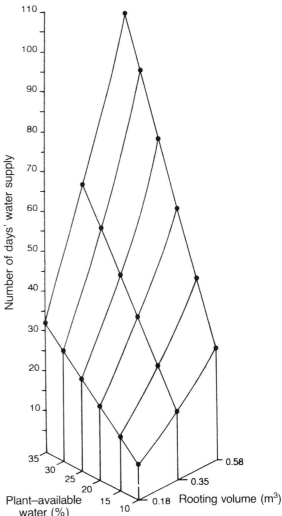

Figure 6.2 *Model of the potential number of days' water supply that a given volume of soil can provide to a newly planted tree, assuming transpiration of 2 litres a day.*

available to the tree, in the absence of rain, is considered to be that within the rooting zone of the tree. As originally proposed, the model assumed a transpiration rate of 2 litres a day. At this rate, the amount of water available to the tree is not likely to last more than 2 or 3 weeks.

Exactly how realistic is a transpiration loss of 2 litres a day was tested by planting standard *Platanus × hispanica* trees with a known root dimension into pots containing 120 litres of a 2:1 peat:sand substrate that had an available water content of 30–35%, the approximate

39

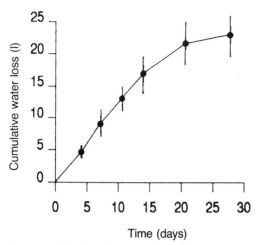

Figure 6.3 *Cumulative water loss from standard* Platanus × hispanica *over 28 days; vertical bars indicate standard errors.*

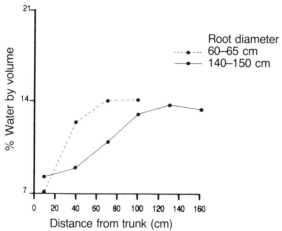

Figure 6.4 *Water tensions that developed around roots of* Platanus × hispanica *after 21 days' drought.*

volume available to the root system would be 120 litres if the trees were planted directly in an unfavourable site. Evaporation from the soil was prevented and the trees were weighed to assess transpirational losses (Figure 6.3).

Whilst the trees remained unstressed (over the first 14 days), transpirational losses amounted to 1.20 litres a day (expressed on a leaf area basis, 11.7 g dm^{-2} day^{-1}), similar to rates reported by other workers (Kramer and Kozlowski, 1979). This simple experiment provides all the values necessary to test the model proposed by Gilbertson *et al.* (1987), as the following values were known: (i) transpirational losses (1.2 litres a day), (ii) volume of soil occupied by the root system (120 litres), (iii) water-holding capacity of the substrate (29–35%) and (iv) number of days' water supply provided by the system (28).

Using these values the model would predict that the system could supply water for about 30–35 days. Thus, the actual value obtained, 28 days, lies within 10% of the lower value predicted by the model. It must be noted that the model is extremely simple and takes no account of edaphic factors, such as changes in hydraulic conduction of water as the soil water potential declines, physiological variables (e.g. closure of stomata) or indeed climatological variables altering the rate of transpiration. Nevertheless, the model appears to provide a good approximation of the situation affecting a newly planted tree.

Increasing the size of the root system transplanted with the tree

The model indicates clearly that the amount of soil water that a newly planted tree can exploit is dependent upon the volume of soil that the root system occupies. A simple increase in the area exploited by the root system from 0.18 m^3 to 0.35 m^3 should increase the potential water supply from 12 to 35 days, assuming that the plant-available water content of the substrate is 15%, typical of many urban soils.

The growth and water tensions that developed around standard *Platanus × hispanica* trees, subjected to drought, planted with large (1.4 m diameter on planting) or a small (0.6 m diameter on planting) root system, more typical of that of many standard trees, were therefore determined. The trees were planted into a polythene tunnel house so that their root systems were protected but their canopies were exposed above the house, to regulate the water input into the system. The water tensions were obtained by using a neutron probe, designed to measure *in situ* the volumetric water content of the soil. The results showed clearly the localised nature of the water available to the tree (Figure 6.4). Trees with larger root systems could be considered to have the potential to exploit a much larger reservoir of water.

Table 6.2 Effect of size of the root system of *Platanus x hispanica* at planting on subsequent shoot extension.

	Diameter of root system at planting (cm)	
	60–65	140–150
Mean shoot extension (cm)	421	1238.25

The trees with the larger root system on planting suffered much less transplant shock, by producing three times more shoot extension than those with smaller root systems (Table 6.2).

Improving water availability

Water availability can obviously be improved by irrigation. It has been shown that irrigating newly planted trees can have dramatic effects on root growth (Table 6.1). Irrigation can, however, be expensive and it can be difficult to ensure that the irrigation water penetrates the soil containing the tree roots, rather than running off. There are methods of overcoming this problem but they require further capital expenditure, such as the use of perforated pipes placed around the root system.

An alternative approach to increasing the amount of water available to the tree is to incorporate, into the soil of the planting pit, materials that have a high water-holding capacity. The most obvious and widely used of these is peat. Recently, however, the use of soil polymers has been suggested. Some cross-linked polyacrylamides have the ability to absorb up to 500 times their own weight of distilled water, although their absorptive capacity is dependent upon the conductivity of the soil solution (Woodhouse, 1989). The results of a field investigation into the merits of using either a peat or Aquastore (a cross-linked polyacrylamide polymer) in the first year of growth of *Acer pseudoplatanus* transplants growing on a sandy loam soil are shown in Figure 6.5. No artificial irrigation was given throughout this experiment.

Amelioration with either peat or Aquastore appeared to be beneficial, leading to an increase in root weight, extension growth and total plant weight and a reduction in the shoot:root ratio. Aquastore performed better than peat.

Such positive benefits of irrigation or soil amelioration in promoting root growth may be quickly eliminated unless weed competition is controlled. Davies (1987) clearly demonstrated that root growth is as sensitive as shoot growth to weed competition. The antagonistic effects of weed growth are clearly related to competition for water (Newton and Preest, 1988). Since grass is one of the most competitive of all the weeds in urban environments, an experiment examining water losses of trees planted with and without grass was carried out. *Acer pseudoplatanus* transplants were planted into 24-litre pots. Water loss from the system was measure by weighing the pots at intervals. This showed clearly that grass can deplete the water reserves of a given volume of soil, twice as quickly as the tree alone (Figure 6.6).

Other factors influencing establishment

Water stress is not the sole problem affecting root growth of newly planted trees, but it is the principal problem and the one that can quickly kill the tree. Although of minor importance during the early stages of establishment, there are other factors that could affect root growth, the most important of which is the nutrient status of the soil and tree.

During the early stages of establishment, the nutrient status of the tree is probably more important in determining root growth than the nutrient status of the substrate. A positive correlation has been shown between the concentration of storage nitrogen in the tree and growth in the following spring (Tromp, 1983). Thus, fertilising the trees whilst still in the nursery could have important benefits on the growth of the tree following transplanting. However, in very poor soils, soil nutrients can have a profound effect on root growth (Kendle, 1988). Other factors, such as soil compaction, soil aeration and waterlogging, are discussed by Hunt *et al.* (1991).

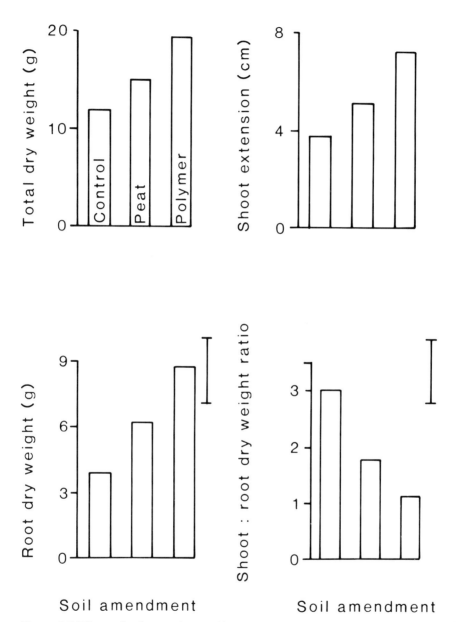

Figure 6.5 *Effects of soil amendment with peat or polymer (Aquastore) on the growth in the field of* Acer pseudoplatanus; *vertical bars indicate L.S.D.*

Conclusion

The experiments reported all emphasise the serious problems likely to arise in newly planted trees because of the severe restrictions to their water supply caused by their truncated root systems. The benefits of planting trees with larger root systems can readily be demonstrated.

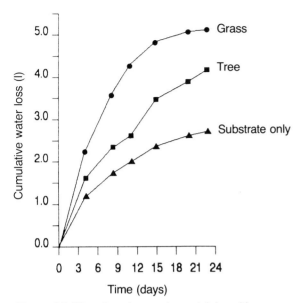

Figure 6.6 *Water loss from pots containing either transplants of* Acer pseudoplatanus *or grass.*

Increasing the size of the root system on transplanting increases the volume of soil water that is readily exploited by the tree; clearly resulting in less transplant shock and increased shoot extension. The optimum size of the root system for different species and stock types is worthy of further investigation.

The situation today, when there is no clear British Standard for the size of the root system (except for semi-mature tree stock), is far from satisfactory. It is imperative that newly planted trees do not face any periods of drought as this will greatly extend the establishment period, and a vicious circle could arise in which, because of the small root system, the water supply is restricted, which, in turn, restricts root growth and, consequently, restricts the volume of water available to the tree the following season.

Soil amelioration with material with a high water-holding capacity, such as peat or certain cross-linked polymers, appears to be an alternative approach, reducing the chances of water stress and promoting root development. It is an area which merits further research.

REFERENCES

DAVIES, R. (1987). *Trees and weeds*. Forestry Commission Handbook 2. HMSO, London.

GILBERTSON, P and BRADSHAW, A.D. (1990). The survival of newly planted trees in inner cities. *Arboricultural Journal* **14**, 287–309.

GILBERTSON, P., KENDLE, A.D. and BRADSHAW, A.D. (1987). Root growth and the problems of growing trees in urban and industrial areas. In *Advances in practical arboriculture*, ed. D. Patch, 59–66. Forestry Commission Bulletin 65. HMSO, London.

HUNT, B., WALMSLEY, T.J. and BRADSHAW, A.D. (1991). Importance of soil physical conditions for urban growth. In *Research for practical arboriculture*, ed. S.J. Hodge, Forestry Commission Bulletin 97. HMSO, London.

KENDLE, A. D. (1988) *The optimisation of tree growth on china clay waste*. Ph.D. thesis, University of Liverpool (unpublished).

KRAMER, P.J. and KOZLOWSKI, T.T. (1979) *Physiology of woody plants*. Academic Press, New York.

NEWTON, M. and PREEST, D.S. (1988). Growth and water relations of Douglas fir (*Pseudotsuga menziesii*) seedlings under different weed control regimes. *Weed Science* **36**, 653–662.

SANDS, R. (1984) Transplanting stress in radiata pine. *Australian Forestry Research* **14**, 67–72.

TROMP, J. (1983). Nutrient reserves in roots of fruit trees, in particular carbohydrates and nitrogen. *Plant and Soil* **71**, 401–413.

WATSON, G.W. (1987) The relationship of root growth and tree vigour following transplanting. *Arboricultural Journal* **11**, 97–104.

WITHERSPOON, W.R. and LUMIS, G.P. (1986). Root generation of *Tilia cordata* in response to root exposure and soil moisture levels. *Journal of Arboriculture* **12**, 165–168.

WOODHOUSE, J.M. (1989). *Water storing polymers as aids to vegetation establishment on arid soils*. Ph.D. thesis, University of Liverpool (unpublished).

Discussion

S.J. Hodge (Forestry Commission)

What is your explanation for differences between Forestry Commission research findings and your own? Did you water your plants?

T.J. Walmsley

I used a better polymer. I watered when the first plant wilted. The polymer I used increased available water to the plants.

A.D. Bradshaw (Liverpool University)

The previous question cannot be left without challenge. The research used plants better than the average used in urban planting. The site was poor, and typical of urban planting sites. Research was carried out carefully and thus site analysis and findings need careful scrutiny for a logical solution to be revealed.

A.S. Kirkham (Royal Botanic Gardens, Kew)

Broadleaf P4 is being used at Kew Gardens with good results. It is only used where soil conditions require it and in conjunction with irrigation. It must be used with care, because, if placed under newly planted trees, it can push the plant out of its pit as it swells.

I. Mobbs (Fountain Forestry)

Manufacturers of Broadleaf P4 also make root dips. Have you had any experience of these and their usefulness?

T. J. Walmsley

Polymers provide a water reserve for the tree, dips are quite different and I have no experience. I have used polymers for some years with unskilled staff and without careful controls and obtained good results.

A. D. Bradshaw (Liverpool University)

There are many polymers on the market and they differ a lot. Research is required to determine which are of use and which are not in different circumstances.

Paper 7

Analysis of performance of semi-mature trees in relation to a high water-table

C.R. Norton, Heriot-Watt University, Riccarton Campus, Currie, Edinburgh, EH14 4AS, U.K.

Summary

Information on tree condition and growth, and depth of soil aeration was collected for 522 trees of six genera at the Glasgow Garden Festival in 1988. Factor analysis was used to indicate the relationships between the parameters assessed. Depth of aerobic soil was closely correlated with several measures of tree performance, indicating that the generally high water-table on the Garden Festival site would limit tree growth.

Introduction

Ageing is a physiological process which is likely to determine the performance of newly transplanted trees. It is the primary reason for the death of woody plants and is distinctly different from secondary infection, e.g. insect or disease damage. Vigorous trees should be less likely to succumb to secondary disorders than those declining in vigour through ageing.

Certain factors which stop growth can seriously impair tree performance and survival, through primary or secondary mechanisms. One of the most severe is transplanting, which necessitates a high proportion of root removal for large specimens.

This paper describes the results of a study of the performance of semi-mature trees at the Glasgow Garden Festival during 1988, to document performance of semi-mature trees after transplanting and determine the main limitations to growth.

Methods

Transect lines were set out at 30 m intervals across the site and data were recorded from 522 trees which grew on or near the lines. The genera sampled were *Acer, Aesculus, Alnus, Pinus, Populus* and *Tilia*.

Data were included on height, girth and extension growth and on tree condition, leaf cover, leaf colour, vigour and dieback. Depth of aerobic soil was recorded using the steel rod method (Carnell and Anderson, 1986). The data were subjected to factor analysis for each genus and Pearson correlation analyses.

Results

For *Aesculus* (Table 7.1), three broad groups of factors accounted for 72% of the total variance: (i) size/root ball, (ii) condition and (iii) previous

Table 7.1 Factor analysis for *Aesculus*.

Variable	Factors		
	1	2	3
Height	0.891		
Girth	0.895		
Condition		0.669	
Leaf cover		0.706	
Leaf colour			0.568
Vigour		0.806	
Dieback		0.735	
Root ball	−0.801		
Wind rock			0.491
Previous growth			0.829
Current growth		0.732	0.525
Radius of crown	0.889		
Variance accounted for (%)	31	25	16

45

Table 7.2 Factor analysis for *Alnus*.

Variable	Factors		
	1	2	3
Height	0.837		
Girth	0.608		
Condition		0.933	
Leaf cover		0.914	
Leaf colour	0.711		
Vigour	0.829		
Dieback	0.863		
Wind rock	0.897		
Previous growth	0.818		
Current growth			0.935
Radius of crown	0.880		
Variance accounted for (%)	45	22	15

Table 7.3 Tree condition over all genera (a total of 522 trees of *Acer, Aesculus, Alnus, Pinus, Populus* and *Tilia*).

Condition	Percentage of trees
Dead	2
Poor	45
Acceptable	45
Good	8

growth. In *Alnus* (Table 7.2), the factors were (i) size/vigour, (ii) condition and (iii) current growth, and these accounted for 82% of the total variation. Similar analyses were conducted for the other genera but are not reported here as the results were similar.

Current and previous year's extension growth was compared across the genera tested. No genus showed a significant increase in extension growth; indeed some showed a decrease in extension growth.

Only 8% of the trees were judged to be in good condition (Table 7.3), while 45% were in an acceptable condition. The remainder were either in poor condition or dead.

Extension growth was generally correlated with condition, there being a strong relationship between low extension growth and poor condition.

Depth of aerobic soil was closely correlated with several measures of tree performance, most notably leaf cover, leaf colour, general condition score and the rate of current and previous season's shoot extension.

Discussion

Performance of semi-mature trees at the Glasgow Garden Festival did not match expectation although, during the Festival, the trees met their objective. Using factor analysis, it was possible to determine which measurements broadly duplicated each other and therefore the most economic form of data collection for future experiments. Data on size, condition and extension growth were the most useful to account for the maximum variability.

In a comparison between the current and the previous year's extension growth, a general reduction in growth, or at least no significant increase in growth, was recorded. The subjective measures suggested that tree performance was not good.

The most reasonable explanation for poor performance on this site was that root growth was restricted by a high water-table. This conclusion is borne out by the general significance of the positive correlation between depth of aerobic soil and plant condition. In many cases, the permanent water-table was above the bottom of the root balls, leading to the effective loss to the plant of the submerged portion of the roots.

It is suggested that a root ball broader and shallower than normal may be appropriate for semi-mature trees for growth in sites with a high water-table. Trees with this type of root ball could be raised on a nursery site with a high water-table. There would be a minor disadvantage in the loss of the dead weight of the deeper root ball for anchorage, but this could be compensated for by appropriate guying of the root ball.

ACKNOWLEDGEMENTS
The author wishes to acknowledge the support of the Scottish Development Agency which funded this work and, in particular, John Lindsay for his enthusiastic support of this

work and for reading the manuscript prior to submission.

REFERENCE

CARNELL, R. and ANDERSON, M.A. (1986). A technique for extensive measurement of soil anaerobism by rusting of steel rods. *Forestry* **59**, 129–140.

Discussion

C.G. Bashford (Colin Bashford Associates)
The semi-mature trees shown on the slides appear no bigger than advanced nursery stock. What size were they on average and do they genuinely constitute semi-mature stock? Rumour has it that planting at the Glasgow Festival was poor, hence a poor success rate.

C. R. Norton
Minimum size was 25 cm girth. I agree that there is some ambiguity over size and definition of semi-mature trees. The main reason for the above-average failure rate at Glasgow was the high water-table creating anaerobic conditions in the bottom half of the root ball.

C. G. Bashford
Will roots grow out of the root ball and the pit and achieve adequate stability without support?

C. R. Norton
Size of the root ball is related to transport costs; foreign importers (Italians) tend to produce small root balls to cut costs. Bigger root balls are better for early stability, but tree roots do achieve stability in time, given that the surrounding soil is of a reasonable structure.

T. P. Marsh (Southampton City Council)
Do you recommend that root-balled trees wrapped in hessian or similar material should be left wrapped or unwrapped when planted? Would Italian root-balled trees benefit from soaking before planting?

C. R. Norton
In my experience, Italian root-balled trees do benefit from soaking before planting. Some hessian materials are restrictive to root growth, others less so. Netting, e.g. chicken wire and chain-link wire, appears to restrict growth and should be removed.

Trees in towns

Paper 8
Importance of soil physical conditions for urban tree growth

B. Hunt, T.J. Walmsley and A.D. Bradshaw, Department of Environmental and Evolutionary Biology, University of Liverpool, Liverpool, L69 3BX, U.K.

Summary

For plants, soil physical conditions are important components of the urban environment. In a series of experiments, soil compaction reduced growth of *Tilia platyphyllos* but this was somewhat ameliorated by the incorporation of additional organic matter in the soil. Localised compaction also reduced tree growth, though distorting the rooting pattern at planting did not. Poor drainage of tree pits, which can be readily measured by steel-rod corrosion, also reduced growth of *Acer platanoides* but this was dependent on the timing of waterlogging.

In a survey of urban situations, weed growth, soil compaction and waterlogging ranked higher than soil chemical factors as determinants of tree growth, although other factors not measured must also be important. Nevertheless, good rates of tree growth will only be achieved if soil physical conditions are taken into account in the design, preparation and amelioration of planting sites.

Introduction

Many factors can determine the growth of urban trees. Some, notably weed growth, bad plant handling and transplanting stress, have received a good deal of attention (Davies, 1987; Insley, 1980; Watson, 1987), and firm guidance can be given regarding their prevention and treatment. There is also good evidence concerning the importance of nutrient supply, but conflicting evidence on the response of trees to fertiliser addition suggests that the situation in urban areas is complex (Capel, 1980). Less evidence exists for the effects of many other factors, despite their citation by several authors as important problems for urban trees. Soil compaction, for instance, has been highlighted as a critical component of the urban environment (Patterson 1976; Ruark et al. 1982), but the link between compaction and amenity tree growth is rarely demonstrated, either experimentally or in urban situations.

In studies of urban tree growth the large number of variables involved and their complex interactions makes it difficult to draw firm conclusions about the causes of poor tree performance. Some factors may be important in causing tree death, whilst others are debilitating rather than fatal. This is often combined with variation in the relative importance of different factors over the life-span of trees and from season to season. Any attempt to rank factors determining tree growth can only be a guideline for focusing attention and allocation of resources; it will not be applicable to all years or all situations.

This paper concentrates on aspects of soil structure and drainage, in two stages. First, the effects of soil compaction and waterlogging on tree growth were tested experimentally. Secondly, their importance was put into perspective by examining a wide range of variables measured in the urban environment, including soil structure and drainage, and studying their relationships with the growth of individual urban trees.

Soil compaction, texture and organic matter content

Soil compaction is a complex phenomenon, con-

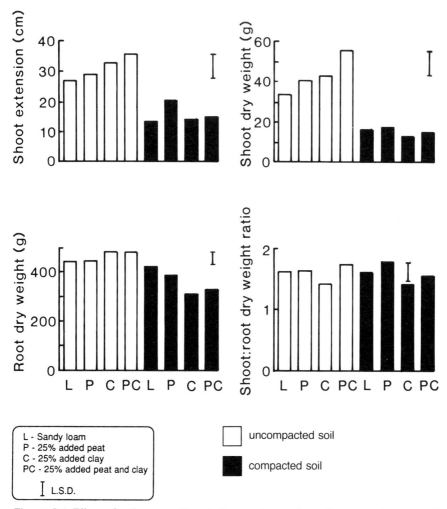

Figure 8.1 *Effect of soil compaction, texture and organic matter content on growth of* Tilia platyphyllos *after 2 years.*

sisting of two main features. First, it leads to greater soil strength, the density of the material increasing so that there is greater cohesiveness, presenting a hard barrier to root penetration. Secondly, there is a reduction in total pore space, as well as a change in the distribution of pore sizes. The number of large (macro) air- and water-filled pores is reduced at the expense of smaller (micro) water-filled pores. Thus, the availability of oxygen to the roots is impaired.

There is likely to be a close link between drainage and compaction as water is more firmly held by micropores, slowing downward percolation and increasing soil saturation time. Compacted soils may, however, also suffer from extremes of water tension, as infiltration rates are reduced, leading to drought during dry periods.

Aeration, soil strength and water-holding properties are all likely to be affected by soil texture and organic matter content. Soils with fine textures tend to have poor structures, with a predominance of micropores, and are easily compacted, especially when wet. Organic materials generally have low densities, are important in the development of good crumb structure, and are likely to increase resistance to compaction. However, a high organic matter content may exacerbate soil anaerobism as micro-organisms break down the material.

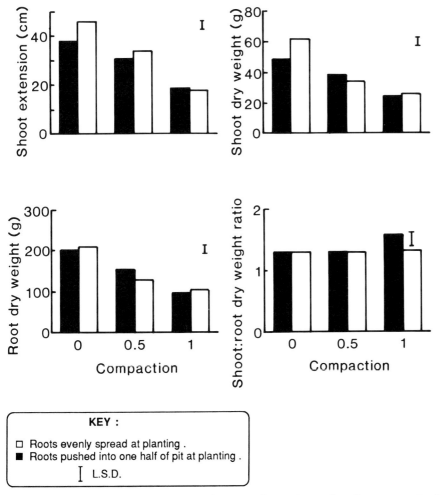

Figure 8.2 *Effect of location and extent of compaction and root distortion on growth of* Tilia platyphyllos *after 2 years.*

To test some of these ideas, an experiment was set up using compacted and uncompacted soils containing 0 or 25% added sedge peat and 0 or 25% added clay, in a full factorial design. The soil materials were mixed in excavated trenches, into which *Tilia platyphyllos* light standards were planted. Compactions were applied every 2 months for 2 years using a trench rammer.

First-year results were inconclusive, possibly because of the overriding effects of transplanting stress. Figure 8.1 shows shoot and root data from the second sampling. The most consistent, statistically significant, trend was the suppression of growth by compaction. Few differences between soil types were observed. The addition of peat to the control (sandy loam) soil, however, improved soil texture so that the reductions in shoot extension and new-shoot weight were not significant. Organic matter appeared not to act as an oxygen sink, but improved resistance to compaction, at least in the amounts added in this experiment, though this pattern is not seen for root weight. Total root weight was a less sensitive measure of tree growth than shoot data, as it was not possible to measure root growth in a single season.

The reduction of growth in the compacted clay treatment was not statistically significant. This treatment was expected to be the most

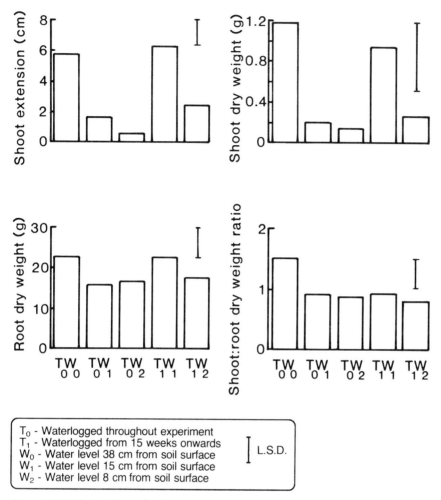

Figure 8.3 *Effects of waterlogging on growth of* Acer platanoides.

severe. Lack of a significant result might be due to poor mixing of the clay and loam, allowing the roots to grow around the clay lenses. When excavated, however, roots could be clearly seen growing through the clay, both in uncompacted and compacted plots. A more likely explanation is that the presence of clay in compacted soil is more critical where overall site drainage is poor; this was not the case in this experiment. Rooting in all soil types was more widespread and fibrous in uncompacted than compacted soils, where roots tended to be shorter and thicker.

Plant response to compaction

Compaction of soil materials may not always be uniform in the urban environment. Trees may adapt to zones of compaction by increased root growth in uncompacted areas. This was clearly observed in the examined root systems of mature roadside trees exposed by the storm of October 1987, but can this occur in younger trees?

In a second experiment, *Tilia platyphyllos* was grown in soils with no compaction, compaction around half of the tree, or compaction all around the tree. Roots were either evenly spread at planting, or pushed into half of the planting pit. The results (Figure 8.2) clearly show that increasing the area of compaction around the tree reduces growth. Distortion of the roots at planting had no significant effect on

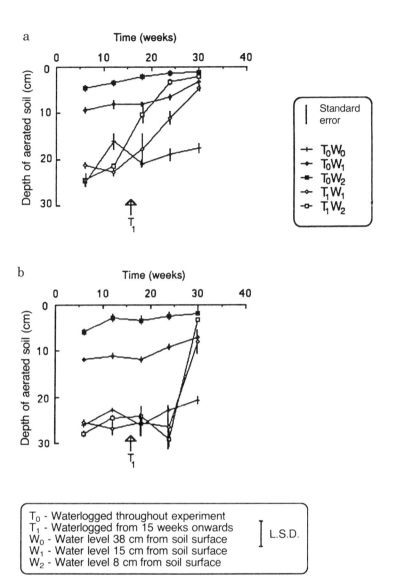

Figure 8.4 Depths of (a) fully and (b) partially aerated soil around waterlogged trees of Acer platanoides.

growth, including root weight, even to trees where the roots were pushed into the compacted half of the rooting zone. Upon excavation, it could be seen that new roots had grown into the uncompacted area from the bole of the root system, compensating for reduced root growth in the compacted soil. Thus, bad planting techniques which distort rooting pattern seem unlikely to affect tree growth adversely during establishment, at least early on, though problems of instability may well occur later.

Drainage

The detrimental effects of poor drainage on forestry crops have been realised for many years (e.g. Green, 1947). In contrast, the importance of good drainage for healthy root development in amenity trees is often overlooked, despite the observation that the root systems of many trees blown over in the storm of October 1987 had clearly been distorted by waterlogged soil horizons.

In a third experiment, the effects of waterlogging were examined. *Acer platanoides* whips were planted into tubs with perforations in the sides to allow water ingress. These were then lowered to different depths into water-filled tanks, creating different degrees of waterlogging: W_0 (38 cm soil to the water surface, control), W_1 (15 cm soil to the water surface) and W_2 (8 cm soil to the water surface). Trees were either waterlogged from the start of the experiment (T_0) or immersed after 15 weeks (T_1), i.e. in mid-May. At the end of October, waterlogging had clearly reduced the shoot growth of the trees (Figure 8.3). This was repeated for root growth, the exception being similar total root weights for the control and T_1W_1 treatment. The control trees consequently had a much higher shoot:root ratio than those in any of the waterlogged treatments. Trees which had been growing in well-aerated soil did not die as soon as they were immersed, perhaps because they had sufficient roots growing near to the surface.

The depth of aerated soil may not be the same as the distance between the soil surface and the water surface. Capillary rise will pull water upwards, and, indeed, water was seen in small depressions at the surface of the most heavily waterlogged trees. There was some fluctuation in the water level as the tanks leaked and were refilled periodically.

To assess this, a technique developed by the Forestry Commission was employed (Carnell and Anderson, 1986). Mild-steel rods were inserted into the tree tubs and removed for measurement and replacement every 6 weeks. The pattern of corrosion and discoloration was recorded as a measure of soil aeration. The length of red-rusted rod corresponds to the depth of fully, permanently aerated soil. Below this is a more indistinct, blackened band of discoloration and corrosion, where the soil is temporarily or partially aerated. This may be due to fluctuations in the water-table and dissolved oxygen.

The most interesting differences in the depths of fully and partially aerated soil were between soils for trees waterlogged in mid-May (Figure 8.4). The red-rusted zone of rod corrosion, i.e. the depth of fully aerated soil, changed soon after immersion. The depth of partially aerated soil, in contrast, increased more gradually, and did not start to level off until September. Thus the depth of full aeration must be a more sensitive indicator of poor drainage than the depth of partial aeration. Although the position of the tree roots in the soil profile was not recorded, making it difficult to judge which measurement is more accurate as an indicator of conditions unsuitable for tree growth, the substantial effects of even partial waterlogging of tree pits means that this technique has great potential in investigations of tree growth.

Survey of trees and soils

Having established that compaction and poor drainage can markedly reduce growth, the critical question is how important are soil conditions and, in particular, physical conditions in determining tree growth in actual urban situations? To answer this, a survey of trees and soils was undertaken. Individual trees from four commonly planted genera (*Tilia*, *Sorbus*, *Platanus* and *Fraxinus*) were selected, growing in grass, paving or gravel-covered ground. A total of 192 trees was examined; they were planted 3–5 years previously and so could have been expected to have recovered from the effects of transplanting stress. This is not to say that the life-histories of trees are unimportant, but the aim of the survey was to examine factors relevant to the growth of the trees after the initial transplanting phase.

Replicate soil samples were collected from the top 20 cm of soil immediately beneath the cover material and bulked for analysis. Sampling was undertaken close to the tree (in the planting pit area) and further away using a large corer. Physical and chemical properties were assessed by normal methods.

Steel rods were inserted into the planting pit, removed for measurement and replaced every 2 months over one year to give a measure of the depth of aerated soil. Only the upper, red-rusted zone was recorded, as an earlier trial in urban soils had shown a complicated pattern of corrosion and discoloration below this level, making divisions between partially, temporarily and

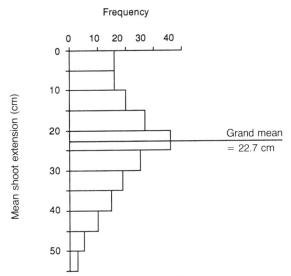

Figure 8.5 *Frequency distribution of mean annual shoot extension for 192 trees (*Tilia, Sorbus, Platanus *and* Fraxinus*).*

non-aerated soil horizons difficult to discern. It is also likely that the patterns of corrosion and discoloration differ over a range of soil types, not simply as a function of soil aeration. Until the technique had been further developed, it was felt prudent to restrict its use to the simplest measurement of pit drainage.

Stem damage was assessed on a scale of 1 to 4 (1 = no damage, 2 = 1-10% trunk girdling, 3 = 10-30% girdling, 4 = 30-100% girdling). For this initial analysis, damage from individual sources, i.e. stakes, ties, guards, vandalism and mowers or strimmers, was summed. The distance from the planting pit to the road kerb was measured, as well as whether the pit was lower than, flush with, or raised above the surrounding ground level.

Experimental sampling techniques are clearly not applicable to surveys of urban tree growth. Visual assessments, such as leaf colour and canopy density, were not used, as there are likely to be errors due to variation between species and cultivars. Girth increment is the measure favoured by foresters, but cannot be applied to urban standard trees because of the stem distortions caused by tree supports (Brown, 1987). Mean shoot extension was therefore adopted as the best measure of growth. This entailed the measurement of a number of leading shoots selected from the ends of major branches in the canopy. Wherever possible, ten shoots were measured; otherwise a minimum of five shoots was used, unless the tree was already dead. Whilst this technique is not wholly satisfactory, including an element of judgement in shoot selection, it is straightforward to record and growth measured by this method is easy to visualise.

There was a wide range in performance of the trees surveyed (Figure 8.5). The grand mean for all genera was 22.7 cm, higher than the 15.5 cm reported by Gilbertson and Bradshaw (1985), but the latter study was of newly planted trees, some of which may have died later and therefore would not be included in the present sample. As shown in Figure 8.5, 25% of trees showed <15 cm mean annual shoot extension and so would take a considerable time to make any impact on the landscape through new growth. However, 25% of trees grew >34 cm so good growth can be achieved.

The first stage of data analysis was to set up a correlation matrix of all the relevant variables. This allowed for the removal of some variables in order to simplify the data set, making subsequent analysis more powerful. Any pair of variables which are very highly correlated for mechanistic reasons represents a 'doubling up' of parameters. One example is total porosity (Pt), which is calculated as a function of bulk density and particle density. As particle density varied little over the range of soils examined, it was not surprising that Pt varied largely as a function of bulk density and was therefore omitted from subsequent analysis.

A simple correlation matrix also allowed the identification of individual variables which show very low probability of correlation with growth (Table 8.1), e.g. available potassium. On the other hand, it also points to factors which might be most important in determining growth, such as stem damage.

Stem damage was highly, positively and significantly correlated with growth over 3 years. As the aim of this paper is to give a summary of the major determinants of urban tree growth in general, variation of the individual components of stem damage is not included (e.g. analysis of

Table 8.1 Survey data: Pearson correlation coefficients for variables against mean shoot extension in trees.

Variable		In planting pit	Outside planting pit
Weed growth		0.242*	
Stem damage		0.286*	
Depth of aerated soil, max.		0.028†	
	min.	0.058	
	mean	0.140*	
Pit sunk, flat or raised		0.071	
Soil bulk density		−0.055	−0.052
Total soil porosity		−0.049	0.048
Organic matter content		−0.041	0.024
Nitrogen, total		0.035†	0.042†
	available	0.071	0.028
	mineralisable	0.029	0.137
Available phosphorus		0.045	−0.032†
Available potassium		−0.035†	−0.081†

Correlation significant at *$P<0.05$, †$P<0.95$

mower or strimmer damage would only be applicable to trees growing in grass). The positive correlation of stem damage with shoot extension, however, implies that growth increases with increasing damage; this requires some explanation. If the components of damage are separated, as in Table 8.2, it becomes clear that most of the correlation of damage with shoot extension is due to that caused by stakes and ties. Trees which are growing well are most likely to put on large girth increments, soon leading to the strangulation and pressure against the stake if trees are not properly cared for. Although damage by these causes may have been classed severe, it may not have impaired shoot extension, though the implications for later growth and stem strength are obvious.

All the major variables were then included in a multivariate analysis. The technique chosen was Principal Components Analysis (PCA), commonly used as a simple tool for exploring areas of importance rather than identifying individual variables. It allows the location of each tree in relation to two of several axes called principal components, consisting of linear combinations of the original variables. Each tree is located by a number which reflects its class of shoot extension over the 3 years of monitoring. Figure 8.6 shows a plot of the first two principal components. If a trend of shoot extension can be observed along a principal component, the trend can be described by the most important variables which make up that component. The outer members of three contrasting shoot extension classes (1, 3 and 5) are marked in Figure 8.6, corresponding to low (Class 1)

Table 8.2 Correlation of different types of stem damage with mean shoot extension of surveyed *Tilia, Sorbus, Platanus* and *Fraxinus*.

Type of damage	Pearson correlation coefficient
Total stem damage	0.286*
Stake damage	0.308*
Tie damage	0.147*
Guard damage	0.030
Mower/strimmer damage	0.003
Vandalism	0.160*

* Significant at $P<0.05$

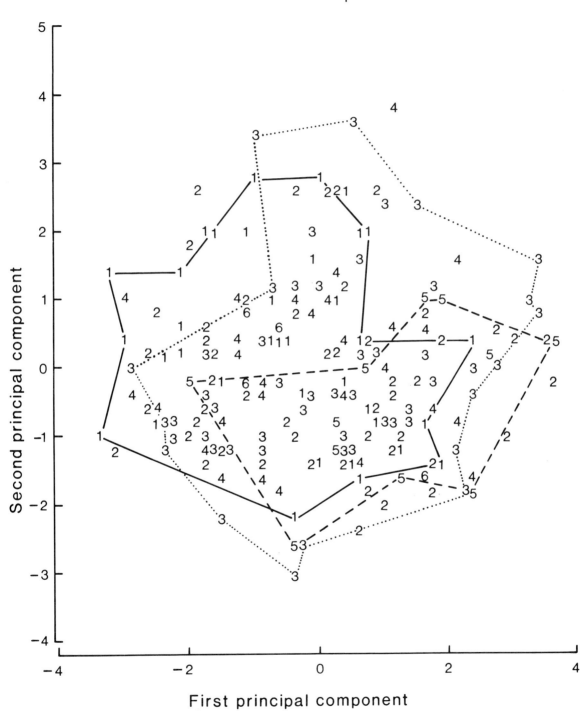

Figure 8.6 Survey of trees (Tilia, Sorbus, Platanus and Fraxinus) and soils: plot of the first two principal components. Mean shoot extension classes 1 = 1–10 cm (——), 2 = 11–20 cm, 3 = 21–30 cm (. . . .), 4 = 31–40 cm, 5 = 41–50 cm (_ _ _ _) and 6 = 51–60 cm.

Table 8.3 Eigenvalues for the first two principal components, and eigenvectors for their constituent variables.

	Eigenvalue	Proportion of variance described (%)
Principal component 1:	3.207	22.9
Principal component 2:	1.903	13.6

	Component 1	Component 2
Weed growth	0.430	−0.177
Bulk density in pit	−0.427	0.072
outside pit	−0.383	0.143
Depth of aerated soil, mean	0.340	0.002
minimum	0.311	0.037
Pit sunk, flat or raised	0.297	−0.128
Mineralisable nitrogen outside pit	0.188	−0.229
Stem damage	0.178	−0.295
Available phosphorus in pit	0.175	0.264
Organic matter content outside pit	0.156	0.430
Available nitrogen in pit	0.155	0.370
outside pit	0.137	0.388
Organic matter content in pit	0.117	0.491
Mineralisable nitrogen in pit	0.147	0.045

intermediate and high rates of growth, respectively. A trend from low to high growth along the first principal component axis can be discerned. No clear pattern emerges against the second principal component.

Table 8.3 shows the most important variables in the first principal component, by the magnitude of the eigenvector: weed growth is the most important single variable, but the majority of the first component is made up of soil physical parameters, in particular, soil bulk density and aeration. Much less of the first component is explained by soil chemical factors.

Weed growth, which in this survey was taken to include annuals growing in otherwise bare areas and grass allowed to remain against trees in grassed areas, has been shown to be a crucial determinant of tree growth (Davies, 1987). Weed species compete vigorously with trees for water and nutrients. Where these are in short supply, even limited weed growth can adversely affect tree performance. Effective weed control measures must be taken in all situations in which weeds can grow close to the tree and clearly matters in the period after the trees might be considered to have become properly established.

Compaction of the soil materials surrounding the tree, as measured by bulk density, ranks high as a determinant of tree growth. Whilst the values obtained by the core method are subject to some distortion by stones, this effect was reduced by using a large corer and collecting replicate samples, and the values obtained cover the range expected. Organic matter might have been expected to have had an effect. Organic matter content had to be measured by loss on ignition; many urban soils contain large quantities of carbonates, which are volatalised at the temperatures used in this technique. This may explain why organic matter does not appear to have influenced tree growth in the clear way that compaction has done.

Drainage of tree pits, measured by soil aeration and the height of the pit relative to its surroundings, also appears significant. Waterlogging can occur even where surface soils are not compacted; the channelling of surface water runoff into sunken tree pits and poor drainage through compacted subsoil limits tree growth.

Some discussion is needed to explain why stem damage, shown as so important in Table 8.1, seems to be so low in the ranking of Table 8.3. Some of the detail of shoot extension values is lost by grouping into classes for PCA plots. This might affect some variables more than others; only extreme damage may be linked to decreasing growth, whilst quite severe damage can occur to trees which are still growing well, for the reasons described earlier. The conflicting components of stem damage may obscure its importance when grouped with other variables in a principal component. The overall positive correlation of shoot extension with stem damage is largely due to ties, stakes and guards; where weed growth is important, it would be expected that mower and strimmer damage would counteract this trend.

Soil physical conditions, particularly compaction and drainage, therefore do not appear to be important in determining urban tree growth. This statement must, however, be carefully qualified. Table 8.3 shows that only 22.9% of the variation contained within the data set is described by the first principal component. This leaves 77.1% of the variance unaccounted for, which can be explained in two ways.

First, many factors beyond the control of this study were not assessed. In selecting the trees for examination it was assumed that they were no longer suffering from transplanting stress. This may not have been true in all cases. In trees which suffer in the first few years after planting, the effects of transplanting stress may be carried over into later years, continuing to debilitate the tree (Walmsley *et al.*, 1991). Secondly, some of the variables which appear to be unimportant in this data analysis may be major determinants of tree growth at certain times and in certain situations. It would clearly be ridiculous, for instance, to suggest that trees do not require a nitrogen supply. Nevertheless, some sites can still be so poor that growth is restricted (Capel, 1980); however, in this study, soil nutrients appear to be relatively unimportant over the sample population of 192 trees.

Conclusions

Many factors can affect the growth of urban trees. In the past, considerable attention has been paid to soil nutrient conditions. Soil physical conditions are also important in tree growth, but are often overlooked in the preparation of urban tree planting sites. Whilst it is important to concentrate on overcoming transplanting stress and ensuring an adequate water supply, it is clear that factors such as soil compaction and waterlogging may become increasingly manifest as the tree subsequently develops in its site. It has to be remembered that soil structure has major effects on the content and movement of water in soils.

The importance of weeds has again been demonstrated. Weed growth, of course, has effects on both physical and chemical soil parameters, but at least it is a problem which can be dealt with relatively easily after planting. Soil physical problems, however, must be understood and treated from the outset. Three main points are relevant here. First, soil materials should be handled correctly to avoid compaction, for instance by not spreading soils when wet and avoiding overworking by heavy machinery. Secondly, remedial treatments to improve drainage and relieve compaction are best employed before trees are planted, e.g. by ripping to break up compacted subsoil layers. Thirdly, good design can be important in ensuring good soil physical conditions. This can be as simple as placing a low kerb stone around the planting pit to prevent channelling of runoff into the rooting zone, but it can also be as elaborate as the construction of a continuous, positively drained planting pit the entire length of a pavement, as used by some American aborists (Patterson, 1985). If soil physical factors are not understood and attended to, tree growth will suffer.

ACKNOWLEDGEMENTS
The authors would like to thank the Department of the Environment, who funded the research described in this paper.

REFERENCES

BROWN, I.R. (1987). Suffering at the stake. In *Advances in practical arboriculture*, ed. D. Patch, 85–90. Forestry Commission Bulletin 65. HMSO, London.

CAPEL, J.A. (1980). *The establishment and growth of trees in urban and industrial areas*. Ph. D. thesis, University of Liverpool (unpublished).

CARNELL, R. and ANDERSON, M. A. (1986). A technique for extensive field measurement of soil anaerobism by rusting of steel rods. *Forestry* **59**, 129–140.

DAVIES, R.J. (1987). *Trees and weeds*. Forestry Commission Handbook 2. HMSO, London.

GILBERTSON, P. and BRADSHAW, A. D. (1985). Tree survival in cities: the extent and nature of the problem. *Arboricultural Journal* **9**, 131–142.

GREEN, W. E. (1947). Effect of water impoundment on tree mortality and growth. *Journal of Forestry* **45**, 118.

INSLEY, H. (1980). Wasting trees? The effects of handling and post-planting maintenance on the survival and growth of amenity trees. *Arboricultural Journal* **4**, 65–73.

PATTERSON, J.C. (1976). Soil compaction and its effects upon urban vegetation. In *Better trees for metropolitan landscapes*, 91–102, USDA, Forest Service General Technical Report NE–22.

PATTERSON, J.C. (1985). Creative site planning alternatives. *Proceedings of the Conference of the Metropolitan Tree Improvement Alliance* **4**, 85–91.

RUARK, G.A., MADER, D. L. and TATTAR, T. A. (1982). The influence of soil compaction and aeration on the root growth and vigour of trees – a literature review. *Arboricultural Journal* **6**, 251–265.

WALMSLEY, T.J., HUNT, B. and BRADSHAW, A. D. (1991). Root growth, water stress and tree establishment. In *Research for practical arboriculture*, ed. S. J. Hodge, Forestry Commission Bulletin 97. HMSO, London.

WATSON, G.W. (1987). The relationship of root growth and tree vigour following transplanting. *Arboricultural Journal* **11**, 97–104.

Discussion

M. Bulfin (ADAS)

Can you confirm that, in compacted soils, root growth is less restricted than shoot growth?

B. Hunt

In clays, roots can be restricted by compaction. There can be no other explanation for decline in tree health where soils have been compacted for whatever reason.

M. Bulfin

Is it possible to determine growth rates for individual species?

B. Hunt

No. There is not enough known about shoot extension in difficult conditions.

M. Bulfin

In which direction is research going? Are soil properties more important than growing stock?

B. Hunt

The long-term effect of differences in soil structure will affect growth, but not kill trees.

D. A. Seaby (Department of Agriculture, Northern Ireland)

What was the diameter of the mild-steel rods and were they cleaned before use?

B. Hunt

Mild-steel rods 40 cm long and 6 mm in diameter were used. In order to remove engineer's oil, they were cleaned before being pushed in by hand.

A.D. Bradshaw (Liverpool University)

No nutrient effect has been shown, presumably because standard back filling contained an adequate nutrient supply, which was available to scavenging root systems. Physical factors of soil are, therefore, more important in the correlation of growth with damage. Ring tests should show periods of slow growth comparable to damage.

B. Hunt

These relations were not apparent in the results because of other interactions.

Paper 9
A study of urban trees

S.M. Colderick and S.J. Hodge, Forestry Commission, Forest Research Station, Alice Holt Lodge, Farnham, Surrey, GU10 4LH, U.K.

Summary

A 2-year study on the influence of soil and site factors on the growth and condition of urban trees (nine species) is described. Specific methods used to evaluate soil bulk density and aeration may be of value to arboriculturists. Within the study trees, correlations with tree growth were stronger with foliar nutrient concentrations than with planting-pit physical properties. More detailed study has been initiated to examine the importance of soil physical condition over the whole rooting zone on the growth and condition of urban trees.

Introduction

In order to improve tree survival and growth it is necessary to understand how urban trees interact with and are constrained by their environment. Research was carried out by the Forestry Commission between 1987 and 1989 under contract to the Department of the Environment to:

a. gather information about the growth rates of the most commonly planted urban tree species over a range of sites,

b. identify, quantify and assess limiting site conditions and relate them to tree growth and

c. investigate and develop methods of assessing soil conditions.

A two-pronged approach was adopted. The first was to carry out an extensive survey, gathering information on growth rates, general condition, age, incidence of damage and siting of the most commonly planted tree species. The second approach was an intensive study on fewer trees over 2 years.

Methods

Extensive survey

Selection of trees

Thirty towns and cities, in Great Britain, with a population of over 30 000 were selected on a random basis, with a weighting given to population. In each location, randomly selected routes were followed and the first 120 trees encountered were surveyed, giving a total of 3600 trees.

Measurements and assessments

Information was recorded for each tree on species, age (subjectively assessed in 10-year categories), mean annual shoot extension (ten shoots per tree were measured from the ends of main branches around the outside of the tree) and general condition score (on a scale of 1, not fulfilling its role in the landscape to 3, fulfilling its role in the landscape).

Intensive study

Selection of trees

Nine locations throughout southern England were chosen and, in each, six to eight sites were selected at which there were four to six trees of the same species and age growing under similar

Table 9.1 Mean annual shoot extension of a total of 273 urban trees in an intensive study over 2 years.

Tree	Species	Shoot extension (cm)
London plane	*Platanus* × *hispanica*	20.1
Norway maple	*Acer platanoides*	12.1
Birch	*Betula* spp.	11.2
Lime	*Tilia* spp.	18.5
Rowan	*Sorbus aucuparia*	14.0
Robinia	*R. pseudoacacia*	20.5
Ash	*Fraxinus excelsior*	16.3
Golden ash	*F. excelsior* 'Jaspidea'	5.7
Whitebeam	*Sorbus aria* and *S. intermedia*	17.0
Grand mean (all species)		15.1

conditions. This was most commonly a row of trees along the side of a road, but also included small groups of trees within car parks, shopping precincts or urban parkland.

Ten of the most commonly planted species between 15 and 30 years old were included in the study, which encompassed 273 trees in total (Table 9.1).

Measurements and assessments

The following information was recorded for each tree.

1. Mean annual shoot extension: ten shoots per tree were measured from the ends of main branches around the outside of the tree, five lateral and five terminal. The mean of 2 years' growth was used in subsequent analysis.

2. Foliar macronutrient concentrations were determined from foliar samples collected in August over 2 years. Assessment of chloride ion concentration was also included in the foliar analysis because high concentrations derived from road salt have been shown to reduce growth or even kill trees (Dobson, 1991).

3. Planting pit soil bulk density. The capacity of a soil profile to absorb and retain water and to allow gaseous exchange to occur is determined by its structure. Compaction of soil reduces the volume of pore spaces in which air and water are present, thus increasing soil bulk density.

Investigations were undertaken to determine the best method of measuring the bulk density of urban soils. Initially a penetrometer, which measures resistance to a probe pushed into the ground, and a soil coring tool, which takes soil samples, were used. However, both proved unsuitable because of the extremely compacted and stony nature of some soils encountered.

A third method was tried whereby a smooth-sided hole with a capacity of approximately 1 litre was carefully dug. All the soil was kept for weighing. The hole was then lined with polythene and filled with water until the water was flush with the surface. By measuring the amount of water, the volume of the hole was determined and, by weighing the soil after oven drying, bulk density was calculated. This basic, robust method proved to be the most practicable.

4. Planting pit soil aeration. In the first year, soil oxygen concentrations were measured using a hollow probe which extracted air from a depth of 0.5 m. Oxygen concentrations were high, nearly all between 14 and 20%, and none below 10%; these concentrations were higher than expected bearing in mind the compacted nature of many of the sites. A possible explanation is that, as the probe was driven into the often compacted stony ground, fissures and gaps occurred around the probe so that when air was

extracted from the soil some atmospheric air was drawn down the outside of the probe and then up through the probe to the oxygen meter. These results were therefore discarded.

Whilst an adequate oxygen concentration of soil air is important for tree health and vigour *per se*, information was also needed on other aspects of soil aeration. Depth and distribution of aeration through the soil profile and its fluctuation through the year have a strong influence on rooting depth (Büsgen and Münch, 1929).

To investigate these aspects of soil aeration a technique developed by Carnell and Anderson (1986) was used. They have shown that the extent of rusting on mild-steel rods driven into the ground can be used to indicate the presence and extent of anaerobic soil conditions and hence the limits to tree rooting. After 3 months in the ground, the oxidised surfaces of the rods were classified in order of decreasing aeration (Plate 9.1):

i. red/brown rust, indicating a well aerated soil;

ii. raised/crusty black, occurring where rusting has started but has been interrupted, or where rust has been knocked off during removal of the rod from the ground;

iii. shiny, indicating the presence of substances (usually polyphenols or oil products) which have protected the rods from rusting. Shiny metal therefore indicates high concentrations of organic residues and the soil can be classed as inhospitable (M. Anderson, unpublished);

iv. smooth black, occurring where limited amounts of oxygen have diffused into a previously anaerobic zone; and

v. matt grey, totally anaerobic conditions.

The steel rod technique was used to assess soil aeration for three 3-month periods in 1988 and 1989: March–May, June–August and September–November. Measurement of soil aeration is less important during the winter because of the intermittent or slow root growth of trees (M. Anderson, unpublished).

Analysis of data

The relative importance and influence of planting pit aeration, bulk density and foliar nutrient concentrations on shoot extension of the five main species in the study (London plane, lime, *Robinia*, Norway maple and whitebeam) were investigated. Two types of analysis were carried out for each species: regression analysis of each factor individually against shoot extension and a stepwise regression analysis which examined the combined effect of the parameters on shoot extension and ranked them in order of importance. By standardising data between species, these two analyses were also carried out using the combined data from all five species.

Plate 9.1 *Steel rods used for soil aeration assessment.*

Top. *Rod showing appearance before insertion into the ground.*

Upper middle. *Rod showing widespread rusting, indicating well-aerated soil. Shiny patches indicate inhospitable soil.*

Lower middle. *Rod from poorly aerated soil; predominantly matt grey with flecks of rust.*

Bottom. *Rod from totally anaerobic soil indicated by matt grey surface.*

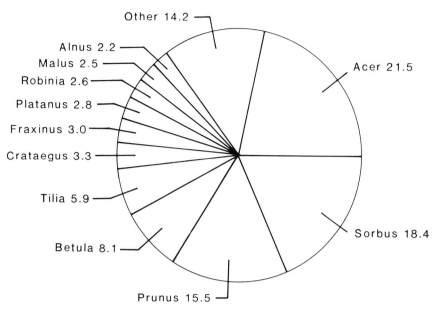

Figure 9.1 *Proportions (%) of different genera studied in the 1989 urban tree survey.*

Results

Extensive survey

Comprehensive results of this survey are presented and discussed by Hodge (1991).

Figure 9.1 shows the proportions of the different genera recorded in the survey. It is striking that five genera make up 69% of the 3600 street trees surveyed, *Acer* making up 21.5% and *Sorbus* 18.4% of the total. The survey revealed that 81% of trees surveyed were less than 35 years old, 11% 35–45 years old and only 8% over 45 years old.

Tree condition was determined by the extent to which the tree was considered to fulfil its role in the landscape; 80% of trees were judged to be fully fulfilling their role. Mean annual shoot extension over all species was 12.4 cm.

Intensive study

Shoot extension

Mean shoot extensions of the species studied are shown in Table 9.1.

Soil bulk density

Soil bulk densities in planting pits ranged between 1.00–1.75 g cm^{-3} (Figure 9.2). Planting pit soil bulk density alone was not significantly correlated with shoot extension in any of the species.

Soil aeration

From the rod assessments, bar charts were produced showing the distribution and extent of anaerobic and inhospitable soil conditions down the 60 cm length of the rod. The bar charts revealed that poorly aerated soil occurred not only at lower depths in the planting pit soil profile, but also close to the surface. On most sites, soil aeration deteriorated appreciably in the autumn, when roots could still be growing. The distribution of anaerobic and inhospitable conditions over three periods is shown in Figure 9.3 for a site with a free-draining soil and in Figure 9.4 for a clay site. The poorly drained and compacted soil showed much less seasonal variation than the free-draining soil.

Before regression analysis could be undertaken, the information yielded by the rods had to be expressed as a single value per tree. To determine whether any particular part of the planting pit soil profile to 60 cm had a greater influence on shoot extension than the others, the rods were assessed in 15-cm quartiles and regression analysis was undertaken by species to determine the percentage variation in shoot

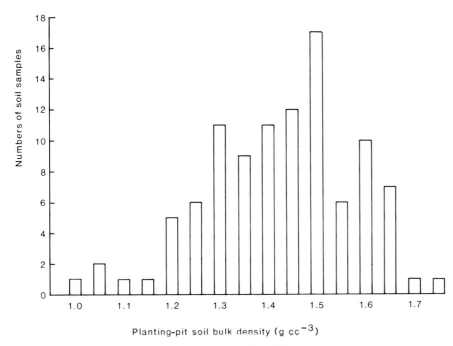

Figure 9.2 *Distribution of soil bulk densities in the 1989 urban tree survey.*

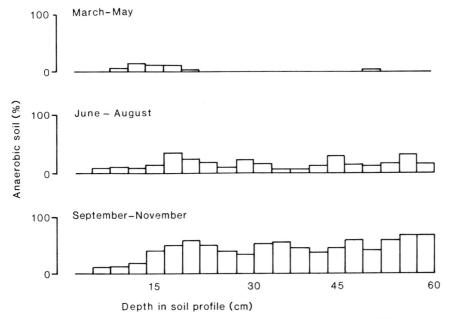

Figure 9.3 *Distribution of anaerobic soil conditions for three periods in 1988 at a site with seasonally waterlogged soil.*

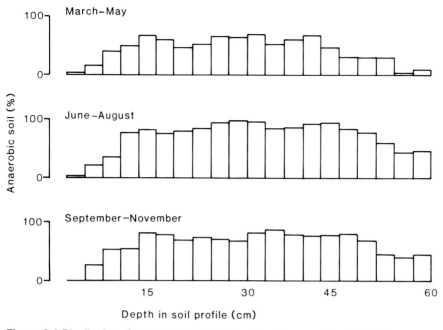

Figure 9.4 *Distribution of anaerobic soil conditions for three periods in 1988 in a compacted clay soil.*

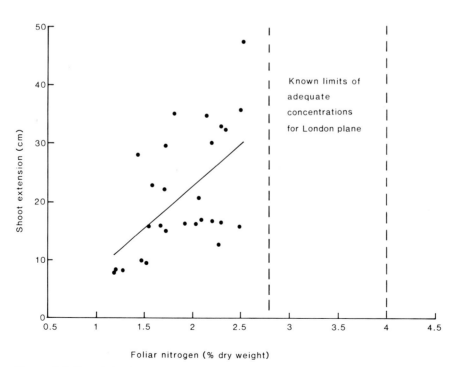

Figure 9.5 *Correlation between foliar nitrogen concentration and shoot extension in London plane (accounts for 30% of variance).*

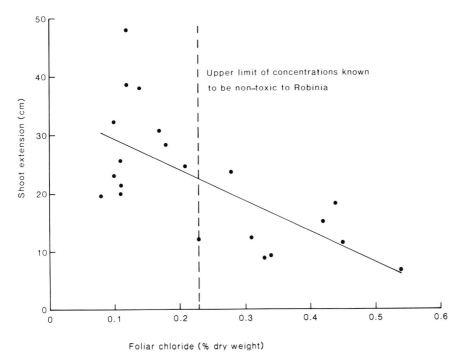

Figure 9.6 *Correlation between foliar chloride concentration and shoot extension in* Robinia *(accounts for 43% of variance).*

extension accounted for by each quartile. In none of the species was there any significant difference between planting pit soil aeration at various depths and shoot extension. For this reason the mean percentage of anaerobic soil over the whole length of each rod was used in subsequent analysis.

To compare the influence on shoot extension of planting pit oxygen status in the three periods, the three sets of data were analysed separately against shoot extension. In none of the species was there any significant correlation between shoot extension and planting pit aeration and there were no significant differences between the correlations with shoot extension in the three periods for any species. The mean of the three sets of data was therefore used in the stepwise analysis.

Foliar nutrients

In general, the trees studied were not deficient in any major nutrients (as defined by van den Burg, 1985), with the exception of whitebeam, which was deficient in magnesium and calcium (as might be expected of a calcicole growing in a range of soil types).

Individual regression analyses showed that shoot extension was correlated positively with foliar concentration of nitrogen in London plane (Figure 9.5), Norway maple (both $P<0.01$) and lime ($P<0.05$); with magnesium ($P<0.001$) and calcium ($P<0.05$) in lime; and with chloride (up to 0.94% of leaf dry weight) in London plane ($P<0.05$); but negatively with chloride concentration in *Robinia* ($P<0.001$) (Figure 9.6).

Stepwise regression analysis of all factors

All soil factors, physical and chemical, were combined in one analysis. As well as identifying significant relationships between soil factors and shoot extension, it also ranked the factors in order of influence on shoot extension (Table 9.2). Planting pit aeration and bulk density were not significantly correlated with shoot extension in any of the species.

Correlations between foliar nutrient concentrations and shoot extension were significant for nitrogen in lime, potassium in London plane,

Table 9.2 Stepwise regression analysis of concentration of foliar nutrients and soil physical factors in the planting pit against shoot extension in five tree species.

Species	N	P	K	Mg	Ca	Cl	Soil bulk density	Soil aeration	Total variation explained (%)
Lime	+*			+*					
% variation	24.7			42.6					67.3
Whitebeam			+		+				
% variation			9.1		4.1				13.2
Norway maple		−		+*	−				
% variation		3.6		21.1	6.7				31.4
Robinia	+				−	−*		+	
% variation	3.3				0.7	41.8		2.3	48.1
London plane	+*		−		+				
% variation	27.6		3.1		9.2				39.9
All species	+**								
% variation	8.7								8.7

+ shows increase and − shows decrease in shoot extension;
* $P<0.05$,   ** $P<0.01$ significance.
Blank spaces indicate analyses that did not yield a correlation with shoot extension.

magnesium in lime and Norway maple, and chloride in *Robinia*.

A stepwise regression analysis of standardised data of all five species together showed that nitrogen was the only correlated factor ($P<0.01$). However, this factor only accounted for 8.7% of the variation in shoot extension in the study as a whole.

There appear, in some instances, to be contradictions between individual regression analyses and stepwise analysis. This is due to the interrelationship between many of the factors assessed. Variation in shoot extension can be well correlated with two or more factors. Whereas regression analysis of individual factors independent of each other attributes variation to each factor in turn, stepwise analysis attributes the relevant amount of variation to the most important factor, and then indicates the factor that best explains the remaining variation. A factor closely related to the one which explained most variation is unlikely to appear important in a stepwise regression as it explains only poorly the residual variation.

Discussion

Soil bulk densities were generally high, with a mean of 1.55 g cm^{-3} (the bulk density of a good agricultural loam is 1.3 g cm^{-3}). Reisinger *et al.* (1988) found that increasing the soil bulk density above 1.4 g cm^{-3} caused a significant decline in root surface area and root mass in loblolly pine (*Pinus taeda*) and that root growth of white oak (*Quercus alba*) was significantly reduced when bulk density increased from 1.0 to 1.5 g cm^{-3}. At the higher bulk densities, low available moisture and impeded aeration would be expected to affect shoot extension, but this appeared not to be so in this study.

Two limitations of the soil sampling techniques used reduce the confidence with which this lack of correlation can be accepted. First, for practical reasons, soil was only sampled to a depth of 20 cm and, secondly, no account was taken of the material outside the planting pit, which would undoubtedly have had major influence on tree growth.

However, the study showed that poorly aerated soil occurs frequently in the top 20 cm of the profile, the zone in which most fine tree roots occur. Where this is the case, nutrients may be present in the soil but unavailable to the tree, and the addition of fertilisers is unlikely to improve growth.

The increase of inhospitable conditions on most sites in autumn is probably due to higher rainfall leading to waterlogging. Soils showing poor aeration through spring, summer and autumn are likely to be badly compacted or impervious. Soils showing high seasonal fluctuations in aeration are likely to be less compacted, but suffering from a seasonally high water table.

In general, soil aeration over any single period showed no stronger correlation with shoot extension than in the others. However, in temperate climates the most critical time of the year for soil oxygen status is generally accepted to be autumn, when rusting of mild-steel rods in other studies has been most strongly correlated with the long-term maximum rooting depth (M. Anderson, unpublished). If this technique is to be more widely adopted as a means of assessing soil aeration, the rods should be inserted during the autumn.

In this study, trees were generally not deficient in any major nutrient (compared with concentrations defined by van den Burg, 1985). Foliar nitrogen had the strongest relationship with shoot extension, high foliar concentrations being significantly correlated with increased shoot extension of London plane ($P<0.01$), Norway maple ($P<0.05$) and lime ($P<0.05$). The higher concentrations of magnesium in *Robinia* ($P<0.05$) and potassium in lime ($P<0.05$) associated with reduced shoot extension could be explained by high concentrations of these ions in the soil limiting uptake of other essential nutrients (Tattar, 1978). The low levels of variation in shoot extension apparent in the study as a whole indicate that other unassessed factors were influencing shoot extension or, as suggested above, that certain soil factors were not assessed adequately. Norway maple, lime and whitebeam showed no significant relationship between foliar chloride concentration and shoot extension and, as expected, the concentrations encountered were below the known toxic thresholds. In *Robinia*, however, where high concentrations of foliar chloride were found, shoot extension was significantly reduced, variation in concentration accounting for 43% of the variation in shoot extension. The apparent positive relationship of foliar chloride concentration and shoot extension in London plane remains unexplained.

Tree growth is not influenced by any soil or site factor in isolation, but by the combined effect of many interrelated factors. For example, high soil compaction can impede aeration, which in turn can lead to nutrients becoming unavailable to the tree.

Eighteen experiments were carried out between 1981–1988 using various weeding, fertiliser and auger treatments to improve the growth of amenity trees. These experiments have established that the greatest diameter growth and foliar nutrient response to fertiliser application or soil aeration by augering came from the use of high-nitrogen fertilisers alone (most of the experiment sites were not particularly compacted and soil augering was found to be more effective as a means of introducing fertiliser into the soil than as a decompaction treatment).

Conclusions

While urban trees generally grew slowly, 80% of trees in the extensive survey were in good condition and fulfilling their role in the landscape.

The intensive study showed that foliar nutrient concentrations were better correlated with shoot extension than were planting pit physical factors. However, limitations to the accuracy of assessment of soil physical characteristics may have resulted in underestimates of their effect on tree growth; soil physical factors do affect root development and nutrient availability, and trees growing in heavily compacted or waterlogged soil are unlikely to respond to fertiliser application.

Of all the factors assessed, nitrogen showed the strongest correlation with shoot extension, indicating that application of nitrogen may be a means of enhancing tree growth and appearance where soil physical factors are not limiting.

Assessment of soil aeration by using steel rods was promising, but more work is required to refine the technique and to verify its value in urban soils. The bulk density of urban soils in the top 20 cm of the soil profile near to trees was generally high, and poorly aerated soil was often found throughout the planting pit soil profile. There were limitations to the methods of assessing physical factors. More development of techniques and extensive assessments over the whole rooting zone are needed before firm conclusions can be drawn.

The analysis and interpretation of results was made extremely difficult by the many variable factors included (and possibly by relationships between factors assessed and factors not included in the analysis). However, the study has shown that tree growth within the urban environment is limited by a variety of factors, none of which was consistently correlated with the shoot extension of the species studied. The complexity of the nature of and relationship between the many and varied soil and site factors influencing urban tree growth and condition requires extensive and detailed research before the factors limiting the growth of urban trees in any particular soil type or planting position can be predicted and amelioration correctly prescribed. More research (funded by the Department of the Environment), currently underway, seeks to relate growth and condition of urban trees to soil and site characteristics over the whole rooting zone.

ACKNOWLEDGEMENTS

The work described was undertaken with funding from the Department of the Environment.

REFERENCES

BURG, J. VAN DEN (1985). *Foliar analysis for determination of tree nutrient status – a compilation of literature data.* Report 414. Rijksinstituut voor Onderzoek in de Bos- en Landschapsbouw 'De Dorschkamp', Wageningen.

BÜSGEN, M. and MÜNCH, E. (1929). *The structure and life of forest trees.* Chapman and Hall, London.

CARNELL, R. and ANDERSON, M. (1986). A technique for extensive field measurement of soil anaerobism by rusting of steel rods. *Forestry* **59**, 129–140.

DOBSON, M.C. (1991). Damage from de-icing salt to trees and shrubs and its amelioration. In *Research for practical arboriculture*, ed. S. J. Hodge, Forestry Commission Bulletin 97. HMSO, London.

GILBERTSON, P. and BRADSHAW, A.D. (1985). Tree survival in cities; the extent and nature of the problem. *Arboricultural Journal* **9**, 131–142.

HODGE, S.J. (1991). *Urban trees – a survey of street trees in England.* Forestry Commission Bulletin 99. HMSO, London.

REISINGER, T.W., SIMMONS, G. L. and POPE, P.E. (1988). The impact of timber harvesting on soil properties, and seedling growth in the south. *Southern Journal of Applied Forestry* **12**, 58–67.

TATTAR, T.A. (1978). *Diseases of shade trees.* Academic Press, New York.

Discussion

D.P. O'Callaghan (Lancashire College of Agriculture and Horticulture)
Is the quality of soil important? How beneficial are nutrient levels to extension growth?

S.M Colderick
Correct. These studies on urban tree growth have necessarily been limited by experimental technique. Nutrient deficiencies may be displayed, but not necessarily by specific elements. It is also suspected that soil compaction affects growth.

B. Hunt (Liverpool University)
If aeration is poor, tree growth is reduced. Anaerobic conditions determine nutrient availability and therefore aeration is related to nutrient supply.

S.M. Colderick
That is true: in these experiments only aeration was measured.

B. Hunt
   Would starch concentration be a better measurement of tree condition than shoot extension?

S.M. Colderick
   Shoot extension is not the perfect measure.

J. C. Peters (Department of the Environment)
   Can you explain the graphs of aerobic conditions?

S.M. Colderick
   They demonstrate oxygen availability and relationship with soil depth. No account was taken of soil temperatures as all trees on all sites were affected.

J. Kopinga ('De Dorschkamp', Wageningen)
   If roots are denied oxygen, will they display stress symptoms?

S.M. Colderick
   Attempts have been made to measure oxygen concentrations, but, so far, techniques have not proved satisfactory.

S.J. Hodge (Forestry Commission)
   Field trials on plane (*Platanus* × *acerifolia*) are to be laid down at Milton Keynes to examine soil aeration, moisture content, tree vitality and extent of rooting of street trees. It is hoped to determine the extent of tree roots under paving.

Paper 10
Benefits of amenity trees

J.R. Matthews, Cobham Resource Consultants, Avalon House, Marcham Road, Abingdon, Oxford, OX14 1UG, U.K.

Summary

As a nation, we spend a lot of money on amenity trees. Research, funded by the Department of the Environment, aims to investigate the objectives of urban tree planting schemes and whether the objectives are being met and to determine the true benefits and costs of tree planting. The study was commissioned in 1989, but the results are not yet available.

Introduction

In late 1989 Cobham Resource Consultants (CRC) were awarded a contract from the Department of the Environment (DoE) to investigate the benefits of amenity trees. It is helpful to examine the background thinking that lead to this research project. In February 1988 the Review Group on Arboriculture presented its report to the Forestry Research Co-ordination Committee (FRCC). The report made eleven main sets of recommendations, but this project is concerned only with those relating to policy and design. The Review Group felt that amenity tree management suffered from a lack of clear objectives and policies and recommended that

> research to gather the information which must form the basis of these policies is essential, especially on the socio-economic benefits of existing trees, on inventory techniques, on where amenity trees should be planted, and on the funding of tree management. A survey of the performance of existing trees is required.

The response of the FRCC was to recommend that the DoE, together with the Economic and Social Research Council consider what socio-economic research could be undertaken to find out where most money is spent on amenity trees in terms of location and management costs. The DoE response was to issue the following brief to consultants in 1989:

> to assess the benefits to society of amenity trees in relation to the costs of providing and maintaining those benefits and to assess the performance of various approaches to tree planting and maintenance in respect of such benefits and their cost-effectiveness.

In undertaking the project the consultant should look at a number of tree planting schemes that have been established for a few years, should cover a range of planting agencies including the private, public and voluntary sectors, should look at a limited number of types of planting sites and finally should examine some unsuccessful schemes.

Specifically, the consultant should identify a selection of schemes and locations, and for each scheme investigate:

> why it was planted and why it was planted in the form chosen;
> who was responsible;
> the benefits and cost of planting;
> the aftercare/maintenance plans and costs,
> the current status of the trees and the outcome of the maintenance regime;
> how far the potential benefits have been realised.

As far as possible all of the benefits should be expressed in monetary terms.

What are the issues?

Before looking in detail at the approach to be

adopted, it is important to review the need for research of this kind. The forestry, agriculture and horticulture industries in the U.K. are well supported by both public and private bodies undertaking research and development. All of these industries produce identifiable products which are either grown in the U.K. or imported. As such, they can be easily costed and their worth to the nation evaluated. Arboriculture is not in this category. Its product, the amenity tree, is particularly hard to value. However, as an industry it is of considerable size. There are estimated to be over 100 million amenity trees in urban areas and over 10 000 people employed in the industry; the 1987 wholesale value ex nursery of field grown and container stock was £121 million.

Publicly and politically, trees for amenity are perceived as a good thing. The immediate response to the two recent storms has been to spend £13 million of public money to assist with replacement planting. The investment in new town landscapes alone runs into hundreds of millions of pounds.

This raises a basic question "Why do we plant amenity trees?" The public has only recently become aware of the role of trees in absorbing the so-called greenhouse gases, but we have been planting amenity trees for centuries. Human interest and attention is focused at both ends of

Table 10.1 Main functions and attributes of woody plants (from Cobham, 1990).

Type of function	Description
Visual	Structures of intrinsic beauty and interest
	Induce a variety of feelings associated with 'verdure' and nature
	Soften or contrast with other landscape elements
	Define spaces
	Direct the eyes and feet of users in accordance with design intentions
	Accommodate changes in level
	Relate buildings to each other and to the site
	Can absorb and encompass human activity or provide shade
	Vertical scale, complements that of urban developments
	Can be bought very large from specialist nurseries for instant impact
	Can provide a substantial mass of vegetation on a site very cheaply in terms of lost groundspace
Social	Screening
	Privacy
	Enclosure
Physical	Demarcate boundaries, areas and features
	Act as barriers to people, stock and vehicles
	Direct pedestrian and vehicular circulation
	Provide shelter from wind, rain and glare
	Have a cooling effect, directly by shading and indirectly through transpiration
	Reduce air pollution through filtration of carbon dioxide, dust, etc. and enrichment with oxygen and moisture
	Provide a low-maintenance landscape 'filler' for low-use, buffer areas
Conservation and recreation	Habitats for wildlife and people
	Backdrop and facilities for sport, play and recreation
	Teaching aids for environmental education
Economic	Commercial, e.g. timber
	Semi-commercial, e.g. country crafts
	Estate and local authority uses
	Non-commercial, e.g. kindling, pea and bean sticks, fruit, leaf mould
	Habitats for game and fungi

the tree life cycle. The public is concerned to see trees planted; what development scheme does not include a tree-planting condition? Equally, there is an almost guaranteed outcry whenever a tree is about to be felled, almost regardless of its condition. In between these two events, by and large, trees are expected to fend for themselves. Table 10.1 lists many of the functions expected of amenity trees.

Amenity trees are planted in many locations and with different objectives. For example, the objectives of planting trees on derelict land are likely to differ greatly from those relating to planting in a suburban street. The type of planting will also differ: on derelict land planting is likely to be in large swathes of woodland, while in the suburbs trees are more likely to be found as individual specimens or in formal avenues. What motivated people to plant trees in these situations, why did they choose trees as opposed to some other form of ground cover or land use? Did the people who planned and designed the schemes have a clear understanding of what they were trying to achieve at the outset and and did they know how the trees would realise their objectives? What benefits did the originators of the schemes hope to achieve? These kinds of questions need to be asked, otherwise planners and designers will proceed on the simple basis that 'trees are a good thing', an approach that sooner or later will be called to account. The study will investigate why people plant trees.

Research approach

Information will be collected in three basic areas: (i) the motivations and objectives of those who plant trees, (ii) the physical success achieved in establishing and growing amenity trees and (iii) the social and economic benefits of amenity trees.

Obviously the basic tasks to be undertaken are to examine trees and to talk to the people responsible for planting them. An early decision was that this study would not talk to the consumers of the amenity, the general public, though it is recognised that it is an important aspect.

In selecting a sample of sites to survey, an approach was required which would allow the study to draw some conclusions about what was happening on a national scale. However, to sample randomly would be extremely expensive and 'grossing up' to national figures would need national statistics that do not exist. Some types of planting location are only found in specific areas (e.g. derelict land and urban rehabilitation projects) which make random sampling difficult.

It is considered that significant information can be gathered about certain areas and authorities and planting locations by adopting a case-study approach. Provided enough areas and authorities are sampled, then an indication of the national picture should be obtained. The picture can be elaborated by assessing the resources devoted to amenity tree planting and maintenance in the different types of planting location and by the different authorities. Four types of study area will provide the locations required: cities, medium-sized towns, rural districts and new towns. The study areas should cover most of England and Wales.

The main variable, apart from location of the study area, is the nature of the planting site. Sample selection has been guided by where most resources appear to be used. Most money seems to be going where there are most people and so the sample contains a predominance of urban sites. There are nine planting locations: derelict land reclamation schemes, urban rehabilitation areas, road-side landscaping, urban and suburban streets, development sites, urban parks, farmland, country parks and rural villages.

Table 10.2 shows the location of the study areas and the distribution of the planting locations amongst them.

The selection of sample locations on the ground presents some interesting problems. In sites selected at random it may be difficult to find out who designed the scheme and to interview them. If designers of a scheme are asked to present sites for inspection they may introduce, albeit unwittingly, bias. However, the study requires a selection of sites chosen randomly and a selection of sites about which the designer can be questioned and so two sets of samples are

Table 10.2 Location of the planting sites and study areas in a study of the benefits of amenity trees.

	Derelict land reclamation	Urban rehabilitation areas	Roadside	Urban and suburban streets	Development sites	Urban parks	Farmland	Country parks	Rural villages
Cities									
Coventry			+	+		+			
Cardiff	+	+						+	
Borough of Hackney		+		+		+			
Newcastle	+	+			+				
Leeds	+	+						+	
Towns									
Wrexham					+		+		+
Northampton			+	+			+		
Exeter					+		+		+
Wigan	+	+						+	
Rural districts									
Macclesfield	+						+		+
Mid-Suffolk			+				+		+
Mendip				+			+		+
New towns									
Milton Keynes			+			+		+	
Peterborough					+			+	
Washington			+	+	+				

proposed. The first will be randomly selected and used principally to provide information about the success or failure of amenity tree planting schemes. The second sample set will be provided by designers and appraised in the same way as the first set and against the designers stated objectives for the site.

For the first set of samples, three location categories are allocated to each area (Table 10.2). Within each category, five sites will be sampled. A further five sites will be sampled at random from any of the nine planting locations to produce a total of 20 sites for each geographic area. A total of 300 sites will be investigated. These locations will be selected randomly with guidance to identify certain types of sites as discussed later.

The second set of samples will be obtained by interviewing the relevant officers in the Local Authorities of the 15 study areas. Information will be sought about their attitudes and objectives regarding amenity tree planting, care and maintenance. Any plan or strategy for the area as a whole and for different districts within the area will be investigated. Direction will be sought on where to find specific types of sites, such as derelict land and urban rehabilitation areas. Names of landscape or forestry firms who have been active in the area over the last five or so years and who have designed and implemented a number of amenity tree planting schemes will be sought. The Local Authority officers will also be asked for guidance on finding voluntary or community groups who have planted amenity trees. The officers will be asked to provide a sample of their own amenity tree schemes. Five of these schemes will be discussed in some detail to determine the objectives of each scheme.

Similar interviews will then be held with the identified landscape or forestry firms and the voluntary or community groups. This process will produce, on a national basis, information about the motivations, objectives, etc. of the three main sectors involved in amenity tree planting and maintenance, and will identify a total of 225 sites (75 in each sector) for which the objectives will have been determined. These sites will be visited and the success or failure of the scheme to meet the designers objectives and whether the stated objectives were reasonable for the site will be appraised. These sites are in addition to the 300 identified earlier.

For all 525 sites identified by the two approaches, information will be collected in three ways. First, 20 trees will be sampled at random and their physical status will be assessed. General information about the site will be recorded. Secondly, the benefits accruing from the scheme will be assessed and the objectives of the designers of the scheme will be appraised. The degree to which these objectives are currently being met or are capable of being met will be assessed. Finally, an overall appraisal will be made of the site, and improvements that could be incorporated will be noted. This appraisal will cover species suitability for the site and design and maintenance of the scheme. Details of the method of data recording are presented on the survey forms (Appendix 10.1).

Measuring economic benefits

All of the approaches rely on an indirect method of placing a monetary value on the tree, as the tree itself is not actually traded in the market place.

As part of the study, a literature review of existing methods to assign economic values to trees will be undertaken. The Helliwell system of evaluating trees (Helliwell, 1967) has much to commend it. The scheme has been accepted in law, is valid and attaches a minimum value to a tree. Trees could be considered to have much greater value and, indeed, large fines have been imposed on developers for removing trees to make way for development. Such figures might be approaching a maximum value for a tree. Other approaches, such as asking people how much they would be prepared to pay for trees or what benefits they associate with trees, are of interest but unlikely to contribute much. Initial literature searches have revealed U.S. studies showing house price increases of up to 5% if there are trees in the picture accompanying the house particulars (Anderson and Cordell, 1985), and reporting quicker patient recovery when trees can be seen from hospital beds. It is a major undertaking to carry out such studies in the U.K. and there are probably strong regional differences.

As the study proceeds, attempts will be made to identify other methods for assessing economic benefits, particularly in relation to residential and commercial property values. A desk study of the costs associated in new town landscapes, will be undertaken. It is hoped to determine the extra cost of achieving a predetermined size of tree and degree of cover using large stock as opposed to smaller trees and shrubs. The time gained by this approach would equate to the conscious or unconscious value placed on the amenity benefit, which in turn will equate to the extra cost involved.

REFERENCES

ANDERSON, L.M. and CORDELL, H.K. (1985). Residential property values improved by landscaping with trees. *Southern Journal of Applied Forestry* **9** (3), 162–166.

COBHAM, R. (ed.) (1990). *Amenity landscape management.* E. & F.N. Spon, London.

HELLIWELL, D.R. (1967). The amenity value of trees and woodlands. *Arboricultural Journal* **1**, 128–131.

Appendix 10.1

Form for recording data on benefits of amenity trees

Sample: 225/300

Geographic area: planting agency: private/public/voluntary

Site name:

Location type: Site code

Planting type	**Material**	**Site factors**
narrow belt < 30 m	transplants	exposed
woodland > ¼ ha	whips	wet
small blocks < ¼ ha	half standard	steep slope
individual trees	standard	poor soil
groups	feathered	made up berm

Support protection		**Surface under trees**
for stability		grass – mown/unmown/grazed
against vandalism		shrub/groundcover
against animals		hedgerow
for shelter		existing woodland
weed control		hard landscape
		bare earth

Estimated age
number of years since planting
0–3   3–6   6–9   10–13   13–16   16–19   20+

Criteria achieved/objectives

Criteria	Have they been met now?			Will they be met in the future?		
	Yes	Partially	No	Yes	Partially	No
	☐	☐	☐	☐	☐	☐
	☐	☐	☐	☐	☐	☐
	☐	☐	☐	☐	☐	☐

Current physical benefits

	good	med.	poor
screening	☐	☐	☐
shading	☐	☐	☐
shelter	☐	☐	☐
filtering	☐	☐	☐
wildlife value	☐	☐	☐
soil conservation	☐	☐	☐
retain age diversity	☐	☐	☐

Landscape benefits

	good	med.	poor
aesthetic (form/colour etc.)	☐	☐	☐
visual significance	☐	☐	☐
degree of fit	☐	☐	☐

Economic benefits

	good	med.	poor
timber value	☐	☐	☐
recreation	☐	☐	☐
impact on areas image	☐	☐	☐
mitigates negative impact	☐	☐	☐

Dis-benefits *(list)*

Physical success – health and growth

Tree Number	1	2	3	4	5	6	7	8	9	10	11	12	13	14	15	16	17	18	19	20
dead (tick if dead)	☐	☐	☐	☐	☐	☐	☐	☐	☐	☐	☐	☐	☐	☐	☐	☐	☐	☐	☐	☐
stake damage	☐	☐	☐	☐	☐	☐	☐	☐	☐	☐	☐	☐	☐	☐	☐	☐	☐	☐	☐	☐
tie too tight	☐	☐	☐	☐	☐	☐	☐	☐	☐	☐	☐	☐	☐	☐	☐	☐	☐	☐	☐	☐
damaged branches	☐	☐	☐	☐	☐	☐	☐	☐	☐	☐	☐	☐	☐	☐	☐	☐	☐	☐	☐	☐
snapped leader	☐	☐	☐	☐	☐	☐	☐	☐	☐	☐	☐	☐	☐	☐	☐	☐	☐	☐	☐	☐
basal damage	☐	☐	☐	☐	☐	☐	☐	☐	☐	☐	☐	☐	☐	☐	☐	☐	☐	☐	☐	☐
epicormic sprouts	☐	☐	☐	☐	☐	☐	☐	☐	☐	☐	☐	☐	☐	☐	☐	☐	☐	☐	☐	☐
animal damage	☐	☐	☐	☐	☐	☐	☐	☐	☐	☐	☐	☐	☐	☐	☐	☐	☐	☐	☐	☐
pest damage	☐	☐	☐	☐	☐	☐	☐	☐	☐	☐	☐	☐	☐	☐	☐	☐	☐	☐	☐	☐
diseased	☐	☐	☐	☐	☐	☐	☐	☐	☐	☐	☐	☐	☐	☐	☐	☐	☐	☐	☐	☐
requires thinning	☐	☐	☐	☐	☐	☐	☐	☐	☐	☐	☐	☐	☐	☐	☐	☐	☐	☐	☐	☐
formative prune	☐	☐	☐	☐	☐	☐	☐	☐	☐	☐	☐	☐	☐	☐	☐	☐	☐	☐	☐	☐
vandalism	☐	☐	☐	☐	☐	☐	☐	☐	☐	☐	☐	☐	☐	☐	☐	☐	☐	☐	☐	☐
stability	☐	☐	☐	☐	☐	☐	☐	☐	☐	☐	☐	☐	☐	☐	☐	☐	☐	☐	☐	☐

(A – negligible   B – significant   C – life threatening)

growth	☐	☐	☐	☐	☐	☐	☐	☐	☐	☐	☐	☐	☐	☐	☐	☐	☐	☐	☐	☐

(A – poor   B – acceptable   C – impressive)

Species code	☐	☐	☐	☐	☐	☐	☐	☐	☐	☐	☐	☐	☐	☐	☐	☐	☐	☐	☐	☐

Appraisal – and what improvements could have been made

Species suitability:

Design:

Maintenance:

Site description

Tree stability

Paper 11
Recent storm damage to trees and woodlands in southern Britain

C. P. Quine, Forestry Commission, Northern Research Station, Roslin, Midlothian, EH25 9SY, U.K.

Summary

The storm of 16 October 1987 caused major damage to trees and woodlands in south-east England. Further damage has been experienced in southern Britain, for example on 25 January 1990. This paper gives some information on the wind climate of Britain, describes the main features of the October and January storms and seeks to explain the differences in damage observed. The intensity of the storm is identified as a dominant factor in determining degree of damage rather than the leafiness of broadleaved trees.

Introduction

The wind climate of Britain, and the occurrence of strong winds, is dominated by the passage of frontal depressions. These form in the Atlantic along the Polar front marking the meeting point between cold northerly and warmer southerly air. The usual, but by no means only, track for depressions is to the north-west of the country, often between the Northern Isles and Iceland. This tends to bring the strongest winds to bear on north-west Britain, and this is reflected in maps of average wind speeds and in the recurrence of strong winds.

Wind speeds at any site are influenced by its location, elevation, surrounding topography and terrain (and measurement height above the ground). Terrain or surface roughness has a particular influence on the turbulence of the airflow; as the surface roughness increases (e.g. from sea to open country, to forest, to urban areas) so the turbulence increases. The roughness causes a reduction in mean wind speed, but the maximum, or gust, is not reduced; ratios of gust to mean therefore increase.

The rarity of certain wind speeds is often defined by reference to return periods, which express the average interval (usually in years) between events of defined magnitude. However, return periods only express the average interval between, or probability of, events and it is possible to have a number of events in rapid succession. Cook, 1985, showed that over a 5000-year notional period, a 1 in 50-year event might not occur for 150 years, but, conversely, four events of this magnitude might appear within 50 years and yet still preserve the average interval. It must also be emphasised that the calculation of the return periods assumes that the observations recorded to date are representative of longer time-spans; if there is a periodicity in the occurrence of events (e.g. windy periods at the end of each century) or a trend of increasing or decreasing windiness (e.g. 'greenhouse effect'), the calculations cease to be meaningful.

In the 40 years prior to 1987, catastrophic tree damage due to strong winds was recorded three times: in 1953 in north-east Scotland, in 1968 in west Scotland and in 1976 in Wales, the Midlands and East Anglia (Quine, 1988). More localised, but nevertheless severe, damage has been experienced on other occasions, for example in the Sheffield area in 1962 (Aanensen, 1965).

The storm of October 1987

The strong winds of 16 October 1987 were the result of a deep depression crossing southern Britain. The track of the depression, from west of Spain to the North Sea via Exeter and Hull, was unusual. The strongest winds, to the south of the depression, were considerably stronger than any previously recorded for this part of Britain (Figure 11.1).

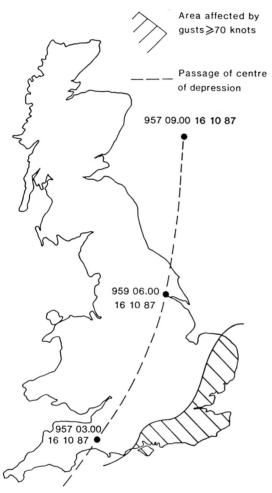

Figure 11.1 *Storm of 16 October 1987 in southern Britain. (Pressure in millibars, time in hours GMT.)*

Gusts of >90 knots were recorded at coastal stations in Kent and Sussex, the highest, 100 knots (115 mph, 51 m s^{-1}), being recorded at Shoreham-by-Sea. A number of inland stations recorded notable gusts, e.g. Herstmonceaux 90 knots, Gatwick 86 knots and Middle Wallop 78 knots. (To convert knots to mph, multiply by 1.15, to m s^{-1} by 0.51 and to km h^{-1} by 1.85.) There was a marked fall off in wind speeds to the north and west, and gusts of >70 knots were not experienced north of a line from Bournemouth/Lymington to Cromer, while gusts of >60 knots were not generally experienced inland north of a line from the Wash to Exeter. Wind speeds recorded in urban areas included 66 knots at Heathrow, 82 knots at the London Weather Centre, 75 knots at the Southampton Weather Centre and 74 knots at the Norwich Weather Centre. Wind directions were mostly 170–220°, i.e. southerly; strong winds are more usually experienced from westerly directions.

The comparison between the measurements from the London Weather Centre (urban site) and Shoeburyness (rural, coastal site) gives an indication of the characteristics of the urban wind climate. The maximum gusts were 82 and 87 knots, respectively, and the hourly means were 40 and 56 knots, giving gust ratios of 2.1 and 1.6. The minimum wind speeds within the hour of maximum gust were 12 and 31 knots, respectively.

The wind speeds were exceptional for low-lying sites in south-east England, although not in absolute British terms; the record gust for a low-level station is 128 knots on 13 February 1989 at Fraserburgh (Roy, 1989).

The rarity of these wind speeds is reflected in the calculated return periods. Return periods of more than 200 years were calculated for stations recording wind speeds of >80 knots, and of 50 years for 70 knots (Burt and Mansfield, 1988). The month of occurrence was also unusual and the return period for a storm of this magnitude in October of any year would be considerably longer than 200 years.

Past records of wind speeds actually experienced, as well as calculated return periods, are relevant in understanding the influence of winds upon the tree population. For example, prior to the gust of 86 knots at Gatwick, the previous highest gust was 70 knots in 1966 and 1962, and the mean annual maximum gust for 1961–1986 was 56 knots (Figure 11.2). In the preceding 20 years there had been little scope for weeding-out aged, unstable trees. Also, in an area of low windiness, the trees may have developed fewer structural adaptations to wind (Jacobs, 1954).

The storm affected some of the most highly wooded counties in Britain, causing major damage to trees and woodlands throughout Kent,

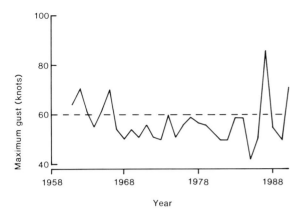

Figure 11.2 *Maximum annual gust recorded at Gatwick Airport, West Sussex; 60 knots threshold line (- - -) indicates approximate wind speed required to cause some tree damage.*

East and West Sussex and Suffolk (Table 11.1) and substantial damage in parts of neighbouring counties. Approximately 15 million trees were windthrown, representing almost 4 million m^3. More coniferous than broadleaved trees were blown over, but the quantity of broadleaved trees was nevertheless very large. More woodland trees were blown over than non-woodland, and many commentators noted the tendency for non-woodland urban trees to be more vulnerable than rural trees. Damage to roots (e.g. by trench digging) or restriction of root spread (e.g. by roads or paths) are possible causes, although structural adaptations of non-woodland urban trees to wind may also be less. Severe funnelling of wind between buildings may also be a factor; in urban areas anemometer sites tend to be on rooftops and tests in wind tunnels have shown that wind speeds in funnelling situations are up to twice that of the rooftop airstream.

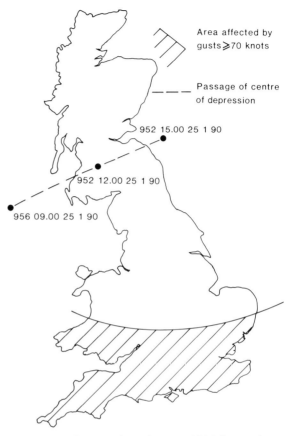

Figure 11.3 *Storm of 25 January 1990 in southern Britain. (Pressure in millibars, time in hours GMT.)*

Table 11.1. Approximate volumes of windthrown timber (m^3 × 10^{-3}) resulting from the storm of 16 October 1987.

	Woodland conifers	Woodland broadleaved trees	Non-woodland conifers and broadleaved trees	Total
West Sussex	350	450	30	830
East Sussex	380	360	10	750
Suffolk	510	200	30	740
Kent	230	310	20	560
Surrey	110	160	20	290
Hampshire	90	110	10	210
Essex	10	90	10	110
Other	230	150	40	420
Total	1910	1830	170	3910

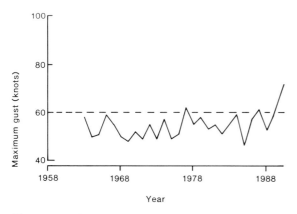

Figure 11.4 *Maximum annual gust recorded at Hurn Airport, Bournemouth, Dorset; 60 knots threshold line indicates approximate wind speed required to cause some tree damage.*

The storm of 25 January 1990

The storm was the result of a deep depression, the track of which crossed Britain close to the England/Scotland border and brought strong winds to a large area of southern Britain (Figure 11.3). The track was similar to that of the depression which caused widespread damage to Wales, East Anglia and the Midlands in January 1976 (Quine, 1989); for both events there is evidence that a trough behind the cold front was important (Shaw *et al.*, 1976). A large area of southern Britain was affected by strong winds during daylight hours, and there was considerable loss of life, frequently from trees falling on vehicles, or people, many of whom were attempting to clear up damage before the storm had abated.

The strongest winds were recorded on coastal sites in south-west Wales and south-west England, for example, gusts of 93 knots at Aberporth and 89 knots at Culdrose. Exceptional winds were also experienced along the south coast (85 knots at Herstmonceaux and 84 knots at Portland Bill, Mount Batten and Shoreham) and throughout the southern counties (74 knots at Lyneham, 75 knots at Farnborough and 72 knots at Stansted), in a broad band to the south of a line from Aberystwyth to Lowestoft. Wind directions were mostly 230–270°, i.e. south-westerly or westerly.

Parts of this area had been affected by gusts of >60 knots in March 1986 and 1987 but many areas, including Dorset and Wiltshire, were affected by winds considerably stronger than experienced previously, e.g. Hurn Airport, Bournemouth (Figure 11.4).

Wind speeds recorded in urban areas included gusts of 79 knots at the Bristol Weather Centre, 76 knots at Heathrow and 75 knots at the London Weather Centre. The Meteorological Office calculated return periods of 25–50 years over much of southern England and at least 50–100 years in the Wiltshire and Hampshire area (Hammond, 1990).

Approximately 1.3 million m^3 of timber was windthrown (about 4 million trees) with damage recorded in many counties of southern England and Wales (Table 11.2). The largest volumes windthrown were in Devon, Hampshire and Dyfed, with other significant quantities in counties such as Dorset, Wiltshire, Avon, and Powys.

Comparison between the 1987 and 1990 storms

In both storms, damage resulted from exceptional winds caused by the passage of deep depressions.

The proportion of broadleaved timber within the estimates of volumes windthrown in the 1990 storm was much lower (12% of woodland timber) than in the 1987 storm (49% of woodland timber), but the composition of high forest within the main affected area was similar in both, i.e. 60% broadleaved forest. This difference might be attributable to the trees being in leaf in 1987 and this has been the subject of much speculation. While not denying the logic of this factor, it is important to emphasise other potential influences. Among these are the history of past damage events in the affected area, the species composition and age-class distribution of the trees and woodlands, the state of management of the crops, the soil types, the preceding rainfall, the intensity of the winds, the direction of the winds and even the type of survey used to derive the damage estimates.

It is not easy to unravel the influence of these factors, and impossible to identify a complete so-

Table 11.2 Approximate volumes of windthrown timber (m³ × 10⁻³) resulting from the storm of 25 January 1990.

	Woodland conifers	Woodland broadleaved trees	Non-woodland (largely broadleaved trees)	Total
Cornwall	26	2	5	33
Devon	209	18	12	239
Somerset	55	4	4	63
Dorset	70	7	5	82
Wiltshire/Avon	71	16	9	96
Hampshire	163	18	5	186
Berkshire	51	8	3	62
Surrey	43	11	8	62
Sussex	41	8	8	57
Oxfordshire	15	16	8	39
Buckinghamshire	7	21	7	35
Dyfed	131	-	-	131
Powys	83	-	-	83
Other England	47	11	10	68
Other Wales	26	-	1	27
Total	1038	140	85	1263

lution. It is, however, possible to illustrate an alternative to the 'leafiness' argument. Figure 11.5a illustrates the low volume of broadleaves windthrown in 1990 compared with 1987, expressed per unit area of high forest in the main affected counties. This appears to confirm the low proportion of broadleaves in the overall damage estimates and supports the importance of leafiness. However, Figure 11.5b shows a similar picture for conifers: the volume blown in 1990 per unit area is also markedly lower than in 1987, yet leafiness (except for larch) cannot be a factor. There are, then, important differences between the two storms in terms of absolute damage as well as in the proportion of broadleaved trees blown over. Is it possible to explain both these differences?

It is known that conifers become vulnerable to

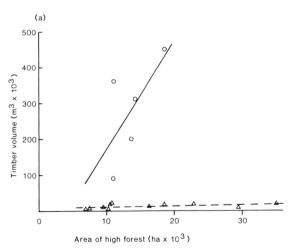 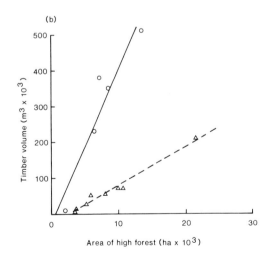

Figure 11.5 Amount of damage (m³ × 10³) compared with area of high forest for each of the main affected counties in southern England for the storms of 16 October 1987 (○) and 25 January 1990 (△) for (a) broadleaved and (b) coniferous trees.

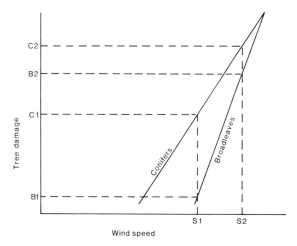

Figure 11.6 *Relationship between wind speed and tree damage, illustrating the possibility of obtaining differences in absolute damage and in the proportions of broadleaved and coniferous trees damaged according to the wind speed of event (e.g. speed S1 → damage B1 + C1; speed S2 → damage B2 + C2).*

windthrow earlier in life than broadleaved trees, that vulnerability of both increases with age and that vulnerability increases with wind speed up to a threshold beyond which significant damage to both is inevitable. This is simplistically portrayed in Figure 11.6. It is possible to identify marked differences in proportions damaged due to small increments of wind speed. If wind speed in the 1990 event lay at the lower end of this spectrum, (S1 in Figure 11.6) and that in the 1987 event lay toward the upper end (S2 in Figure 11.6), this would explain the difference in proportions of conifers and broadleaved trees damaged and the absolute damage levels.

Is there any evidence to support this view and does it conform to estimates from other catastrophic events? There are major differences between the events in the percentage of growing stock damaged within affected counties, e.g. 1–3% for 1990, 13–24% in the main area for 1987, 10–25% in the main area of north-east Scotland in 1953 (Penistan, 1954) and about 15–30% in West Scotland in 1968 (MacKenzie and Martin, 1971). Table 11.3 summarises data on these storms. The proportion of broadleaved timber within the damage total broadly mirrored that of the growing stock in both the 1953 and the 1987 storms (Quine, 1988); data are not precise enough to allow comparisons for the other storms.

The distinguishing feature between the storms appears to be the intensity of wind damage to 10–30% of the total growing stock being associated with gusts of >80 knots. There appears to be a very rapid fall in the proportion of growing stock damaged with reduction in wind speed, so that storms with gusts of about 75 knots only generate damage of <5% overall. Within this latter class, there are likely to be localised areas affected by gusts of >80 knots where much damage will be experienced; in the more intense events these 'hot spots' coalesce to give the greater total damage.

It is therefore possible to explain much of the difference in total degree of damage and in the coniferous and broadleaved proportions of damage by reference to storm intensity, without in-

Table 11.3 Summary of data for catastrophic storms affecting Britain since 1945.

Date of storm	Area affected by 70-knot gusts (km^2)	Max. gust recorded (knots)	Mean of max. gusts recorded within 70-knot zone (knots)	Volume of windthrown timber (m^3 × 10^{-6})	Growing stock windthrown (%)
31 January 1953	370	98	84	1.80	9.7–25.3
15 January 1968	510	102	83	1.64	15–30*†
2 January 1976	890	91	76	0.96	<5
16 October 1987	220	100	80	3.91*	13–24
25 January 1990	690	93	76	1.26*	1–3

* Known to include non-woodland trees.
*† Percentage of crops aged 31 years and over.

voking the leafiness argument. This does not discount the effect of leafiness but emphasises the need for caution in attributing dominant cause to effect. It can be argued that leafiness may be more important for free-standing trees rather than woodland trees, as the increase in drag is not complicated by the potential for reduced penetration of gusts into the woodland canopy. Resolution of this debate must await further fundamental research.

REFERENCES

AANENSEN, C.J.M. (1965). *Gales in Yorkshire in February 1962.* Meteorological Office Geophysical Memoir No. 108, HMSO, London.

BURT, S.D. and MANSFIELD, D.A. (1988). The great storm of 15–16 October 1987. *Weather* **43**, 90–114.

COOK, N.J. (1985). The designer's guide to wind loading of building structures – (*Building Research Establishment Report*) Background, damage survey, wind data and structural classification. Butterworths, Sevenoaks.

HAMMOND, J.M. (1990). Storm in a teacup or winds of change? *Weather* **5**, 443–448.

JACOBS, M.R. (1954). The effect of wind sway on the form and development of *Pinus radiata. Australian Journal of Botany* **2**, 35–51.

MACKENZIE, A.M. and MARTIN, R.K. (1971). Damage to the forests: quantitative statement of areas and volumes of timber windthrown. In *Windblow of Scottish forests in January 1968*, ed. B.W. Holtam, 8–13, Forestry Commission Bulletin 45. HMSO, London.

PENISTAN, M.J. (1954). *The gale of 31 January 1953.* Forestry Commission Report, unpublished.

QUINE, C.P. (1988). Damage to trees and woodlands in the storm of 15–16 October 1987. *Weather* **43**, 114–118.

QUINE, C.P. (1989). Description of the storm and comparison with other storms. In *The 1987 storm: impact and responses,* ed. A.J. Grayson, 3–8, Forestry Commission Bulletin 87. HMSO, London.

ROY, M. (1989). Hazardous weather in Scotland. *Weather* **44**, 220.

SHAW, M.S., HOPKINS, J.S. and CATON, P.G.F. (1976). The gales of 2 January 1976. *Weather* **31**, 172–183.

Discussion

P. White (London Borough of Ealing)
There is not a clear difference between the effects of wind on broadleaved and coniferous trees. However, the difference between conifers in leaf and broadleaves in leaf must be relevant; and conifer roots, remaining active all year, provide anchorage.

C.P. Quine
Branch and stem structure is fundamentally different between conifers and broadleaves, so one cannot simply latch on to one feature, such as leaves versus no leaves. Furthermore, experience of wind exposure in a tree, leads to adaptation of rooting and anchorage.

W. Cathcart (Royal Parks)
It was noted that damage varied with species, e.g. thorns were blown down, but *Sequoia* was not damaged.

C.P. Quine
Yes, *Sequoia* was notable as being 'windfirm', possibly because of stem taper (wide base, narrow upper crown). Larch was also notable because it was blown over when leafless!

Paper 12
Survey of parkland trees in southern England after the gale of October 1987

J. N. Gibbs, Forestry Commission, Forest Research Station, Alice Holt Lodge, Farnham, Surrey, GU10 4LH, U.K.

Summary

In the immediate aftermath of the October 1987 gale, the Pathology Branch of the Forestry Commission, with financial support from the Department of the Environment, surveyed 20 parkland sites in a region just to the west of the worst-affected area in the south-east of England. Wind speeds in this area were generally in the region of 70 knots. A summary of the results is presented here; full details of procedures and results can be found in Gibbs and Greig (1990).

Survey

The objectives were to determine the degree and nature of the damage suffered by different tree species and to identify factors that might have influenced events. Because of the speed with which clearance and remedial work was conducted, the first part of the survey concentrated on trees suffering significant damage. These obviously comprised snapped and uprooted trees, but also included any tree that had lost a branch with a diameter of >30 cm or one-quarter of the tree's diameter at breast height. The second

Table 12.1 Data on incidence of damage on some of the more common taxa encountered in a survey of trees in 20 parkland sites in Britain after the gale in October 1987.

Species	Number of trees	Percentage of damaged trees	Ranking order*
Acer platanoides	132	15	1
Acer pseudoplatanus	185	15	9
Aesculus hippocastanum	411	28	18
Aesculus × carnea	45	38	17
Castanea sativa	104	19	12
Cedrus atlantica	81	12	5
Cedrus other	44	30	3
Fagus sylvatica	442	28	10
Fraxinus excelsior	68	21	14
Pinus nigra	61	18	14
Pinus sylvestris	55	15	4
Platanus × acerifolia	230	11	11
Populus spp.	76	39	19
Quercus cerris	78	25	8
Quercus robur	539	35	16
Quercus rubra	41	17	7
Quercus other	123	28	13
Tilia cordata	104	33	2
Tilia platyphyllos	155	42	20
Tilia vulgaris	532	16	5

* 1 = least damaged

stage of the survey covered the residual 'not significantly affected' trees on each site.

Results and discussion

Full details are given in Gibbs and Greig (1990). Information was collected on 3954 trees, 26% of which were significantly damaged. The failure rate for some of the more common taxa is shown in Table 12.1. Overall percentage data can be misleading as they may be much influenced by data from only one or two plots. In an attempt to overcome this problem a ranking order was devised based on a plot-by-plot comparison of the relative performance of the species. This ranking order is also shown in Table 12.1; for example, *Tilia cordata* is shown as much less damaged in the ranking score than in the percentage data, but the opposite is true for *Platanus* × *acerifolia*. There was a statistical basis for the finding that *Tilia vulgaris*, *Fagus sylvatica*, *Aesculus hippocastanum*, *Quercus robur* and *Tilia platyphyllos* lay on a progression from least to most damaged.

The effect of tree size was considered in relation to height and diameter at breast height. Within some species there was a trend towards greater damage with increasing size, but rarely was this effect statistically significant.

The resistance of the crown of a parkland tree to a gale depends upon many factors, such as size, shape and density. From the data collected it was not possible to produce a significantly precise measure of these features for comparison between species to be worthwhile. However, it was feasible to consider the possible importance of some of these factors within the most commonly encountered species. Crown density was of special interest since there had been some suggestion that, because the gale came while the trees were in leaf, well-formed vigorous trees might have suffered more than decrepit trees with poor crowns. The results showed that this was probably true for *Q. robur* but not for *A. hippocastanum*, *F. sylvatica*, *T. platyphyllos* or *T. vulgaris*.

A total of 1143 failure points was recorded on the 1021 significantly damaged trees. Four failure locations were recognised. Root failure accounted for 431 trees being blown over (38% of the total). This high figure was almost certainly in part due to the heavy rain in the weeks before the gale; by saturating the soil, rain had reduced the shear strength of the soil and hence its ability to provide anchorage for tree roots. 'Main-stem' failures (112 in total) involved the trunks of single-stemmed trees, while 'multi-stem' failure (134) was recorded where the damage affected a vertically ascending stem on a tree with two or more such stems. The remaining 466 failures involved branches: to qualify, these had to have a diameter of 30 cm at the failure point, or exceed a quarter of the tree diameter at breast height. In total, 32% of the failures were associated with appreciable decay, defined as decayed wood occupying at least one-third of the tissue at the failure point.

There were some quite marked differences between taxa. Thus, *Fagus sylvatica*, *Tilia vulgaris* and *T. platyphyllos* commonly failed by being blown over at the roots, while *Aesculus hippocastanum* and *Quercus robur* typically showed branch failure. Appreciable decay was much more common in beech with root failure than in the two limes. Decay was rarely associated with branch failure in *A. hippocastanum* but was quite common in *Q. robur*.

The cause of the decay was determined in some cases from the characteristics of the rot, in others by culturing the fungus in the laboratory. *Laetiporus sulphureus*, 'the chicken of the woods', was the most common cause, being found principally in association with decay in branches. Honey fungus (*Armillaria* spp.) was the next most common and was found on a wide range of hosts, but on only 31 of the 431 wind-thrown trees, thus refuting a view, expressed shortly after the gale, that the wind had principally overthrown trees previously weakened by this fungus.

In the literature on the detection of hazardous trees, attention is drawn to types of structural defect that should be identified when trees are assessed for safety. It was therefore of obvious interest to determine how frequently failure occurred at such points. Some type of structural weakness was associated with 50% of the main-stem failures, 70% of the multi-stem failures and 33% of the branch failures. With multi-stems, over half the failures occurred at weak forks. This is a high incidence but is not perhaps surprising when it is considered that the conditions under which two or more vertically ascending stems develop on a tree are likely to be conducive to the formation of weak forks at their points of origin. Weak forks were also quite commonly associated with failure in branches. In both cases there was rarely external evidence of decay.

Old pruning wounds were associated with 11% of the failure points in the branches and there was usually external evidence of decay, which was also noted in most of the few main-stem and multi-stem failures that involved old pruning wounds. Bark defects were fairly commonly associated with main-stem failure, comprising about 30% of all these failures. Here, and with failure related to bark defects in branches and multi-stems, signs of decay were often visible externally.

In considering failure in relation to structural weakness, it is of obvious interest to consider similar features on trees which did not show significant damage in the gale. Some measure of this was obtained by recording the single most conspicuous 'non-failure point' on these trees. Evaluation of these data show that the occurrence of bark defects on the main stem featured prominently. While such data must be treated with caution – the eye of the observer being more readily attracted to a feature low on the tree than to one high in the branches – it seems possible that such cankers, especially basal cankers buttressed by rolls of callus, are not such a serious hazard feature as their appearance might suggest.

Finally, some consideration must be given to the information collected on soil features. Although the 20 sites spanned a wide range in geological terms, most of the trees were on brown earths. However, there were enough trees on ground water gleys for comparison to be made between these two soil types. From the 'all tree' data, it was determined that the proportion of root failures to total failures was higher on brown earths than on gleys; this effect was shown separately by *Aesculus hippocastanum*, *Fagus sylvatica* and *Quercus robur*. The distribution of the trees in relation to the soils did not, however, permit this effect to be tested statistically. Less surprisingly, for the 'all tree' data there was some evidence for a progressive decrease from shallower to deeper soils in the proportion of failures that involved wind throw.

REFERENCE
GIBBS, J.N. and GREIG, B.J.W. (1990). Survey of parkland trees after the great storm of October 16, 1987. *Arboricultural Journal* **14,** 321–347.

Discussion

C.G. Bashford (Colin Bashford Associates)
Crown density is probably greatest in horse chestnut and beech, both species noted for 'summer branch drop'. Is there a correlation between crown density, site type and wind firmness and should we be paying more attention to the sites in which we plant certain species?

J.N. Gibbs
There is no evidence for saying that either was correlated to crown density.

S.J. Hodge (Forestry Commission)
In the parkland trees that you examined, was the sample of trees with *Armillaria* proportional to the population of *Armillaria*-affected trees as a whole?

J.N. Gibbs
Not proven.

B.J.W. Greig (Forestry Commission)
Some press reports suggested that all the damaged trees were subject to *Armillaria*.

C.L.A. Davis (Department of the Environment)
Where branch failure occurred there was a high incidence of *Laetiporus*. Could this be related to management practices, e.g. flush-cutting?

J.N. Gibbs
It could be. In any case we should not black-ball oak on poor performance which may be management related.

I. Mobbs (Fountain Forestry)
Did storm damage indicate any implications for preventative tree surgery, e.g. crown reduction?

J.N. Gibbs
No. I would hesitate to suggest this as we could not determine what crown treatment would be required for any particular tree in any particular situation.

Paper 13

Tree stability

H.J. Bell, A.R. Dawson, C.J. Baker and C.J. Wright*, Department of Civil Engineering (*Department of Agriculture and Horticulture), University of Nottingham, Nottingham, NG7 2RD, U.K.

Summary

Recent severe storms which caused extensive damage in the south of England have focused attention on tree stability. Whilst the strong winds experienced in 1987, 1989 and early 1990 are not common, they are certainly not freaks. However, trees are blown over more regularly than the frequency of these storms, though fortunately not in such numbers. In this paper, influences on selection mechanisms involved in tree fall and the forces required to produce these effects are considered. The issues are complex, involving the entire tree and its interaction with and response to climatic conditions, the aerial and ground environments and any changes which occur. Differences are observed between species and are further complicated by the biological variability of individual specimens.

Introduction

What goes up, must come down! Trees are not immortal and so inevitably will fall. Tree fall, particularly in high winds, has disrupted transport, power and communication systems and caused structural damage to buildings, injuries to people and even fatalities.

The problem would not be as great if each tree was guaranteed to reach maturity and therefore had a predictable life span. A management regime could be introduced to cull older trees before they become hazardous. However, external influences, such as strong winds, prolonged drought, waterlogging and pests and diseases, increase the incidence of unpredictable premature death.

Current procedures for tree safety and stability in public areas, streets and council-owned property may involve regular inspections by qualified arboriculturists or foresters, possibly with the aid of computerised tree database and maintenance records. However, they may rely on reports from worried members of the public. On privately owned land even the most obvious signs of ill health and weakness in trees may be missed through ignorance. It is not only diseased and dying trees that lose branches, suffer stem breakage or are uprooted; indeed, dying trees may survive storms better than healthy ones. There are no reliable methods of assessing tree stability, although there are methods for investigating the extent of internal decay.

Research at the University of Nottingham

Tree structure and its interaction with and response to wind is being studied in 'isolated' (growing not in woodland but singly, in streets or avenues or in small groups) broadleaved trees in urban or roadside situations. The behaviour of conifers in wind has been studied for many years by the Forestry Commission and other forestry organisations trying to increase and improve timber production in increasingly exposed and unfavourable areas by different planting and management regimes. There has been less research on the response of broadleaved trees to wind, although much of the work on conifers applies to broadleaved trees. The trees that are close to and a part of the urban environment are mostly broadleaved and conflict is bound to occur between their desirability from an amenity point of view and public concern regarding their safety.

Research at Nottingham University is being carried out within the Departments of Civil Engineering and Agriculture and Horticulture, funded by the Science and Engineering Research Council. The purpose of the project is to determine the factors which contribute to tree stability in high winds, to understand those factors, and to outline methods for detecting trees at risk. The following objectives were proposed:

a. to determine the species and specimens which are most susceptible to uprooting in wind,
b. to determine the location of trees most likely to fall, and
c. to investigate the mechanisms of stability and uprooting.

A literature review uncovered much work on the subject of tree stability and highlighted aspects, particularly in broadleaved trees, where knowledge is lacking. Records of fallen trees, especially in urban areas, have not been readily available, the main priority being to remove fallen trees quickly to minimise inconvenience. However, following the strong winds of 16 October 1987, data were collected by the Forestry Commission on 20 parkland sites in the south-east of England (Gibbs, 1991). A survey of tree roots was carried out by Task Force Trees (Countryside Commission) and the Royal Botanic Gardens at Kew (Cutler et al., 1989).

Both sources have kindly permitted access to these data, which are to undergo further analysis, with more data collected by us in Aberdeen following the strong winds on 13 February 1989 and in Devon, Cornwall and Nottinghamshire following recent (spring 1990) gales. It is hoped that statistical analysis will show connections between tree stability and one or more of the fundamental factors illustrated in Figure 13.1.

Aspects of tree stability

Tree stability is mechanically complex and further complicated by biological variability. All parts of the tree, below and above ground, are involved and the immediate environment, i.e. the soil and ground condition, is important. The aerial environment (in terms of exposure, surface roughness and large nearby structures which may alter the force and path of the wind) also affects the tree (Figure 13.1).

Trees have adopted various forms during their evolution, suited to the differing biological and physical conditions encountered in their natural habitats. Each individual specimen is able to adapt to a certain degree in response to changes in its specific environment. Generally, trees adopt a form to maximise photosynthate production within their structural capabilities, i.e. they grow to withstand harsh conditions though not necessarily with a sufficient safety margin to withstand a 50- or 100-year return wind speed (a prerequisite for an engineering structure).

In very exposed areas with frequent strong winds, few trees grow. Those that do may be stunted and have a thicker trunk than normal and examination reveals the growth of 'reaction' wood. In conifers this is compression wood formed on the underside of branches and the leeward side of the trunk. The fibres in the wood have thicker walls and are denser than those in normal wood. The wood is strengthened to withstand compressive forces and contains more lignin and less cellulose than normal. In broadleaved trees, similar wood is formed in tension on the windward side of the tree and the upper side of the branches (Cannell and Coutts, 1988). Estimates of wind speed and direction may be made by measuring the thickness of this reaction wood, providing the terrain is not too steep (Robertson, 1987). Buttressing of the roots is another mechanism adopted by certain tree species to improve stability.

Tree shape and wind interception

The categories conifer and broadleaf immediately create generalised pictures in the mind of a triangular conifer and a square or circular broadleaf (Figure 13.2).

Wind may be a major influence on these shapes, broadleaved trees often affording a larger sail area with the wind forces acting on the canopy at a greater height, i.e. a higher

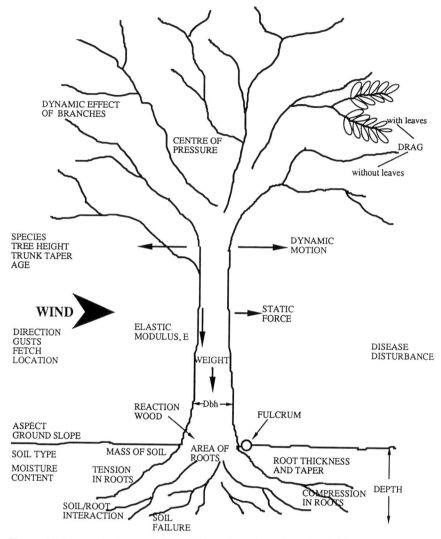

Figure 13.1 Aspects of tree stability (Dbh = diameter at breast height).

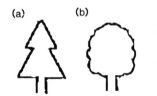

Figure 13.2 Generally perceived images of (a) coniferous and (b) broadleaved tree shapes.

centre of pressure, increasing the turning moment at the base of the tree. This is permissible in the majority of broadleaves, as, being deciduous and losing leaves, they reduce wind interception during the winter when high winds are most frequent. Evergreen broadleaves must find other survival mechanisms, possibly by having smaller leaves, being more streamlined, or having a more compact growth habit.

Observations of tree behaviour in wind reveal that, generally, conifers are more streamlined with the air flow, because of the suppleness of their stems and branches. Young's modulus of elasticity (E) along the grain of the wood is a measure of resistance to deflection or the effective rigidity, large values of E indicating a rigid material. This is a difficult parameter to determine in living wood, but values are available for air-dried and green timber (Cannell and

Morgan, 1987). Values of 7.6 and 8.9 GPa have been obtained for coniferous and broadleaved green timber (soft and hardwood respectively) and values are expected to be lower in living timber which includes bark.

The force exerted by the wind on a structure in the wind direction is known as the 'drag'. For a building, the drag (D) is directly proportional to the square of the wind velocity (V):

$$D = C_D \tfrac{1}{2} \rho A V^2$$

where ρ is the air density, A is the building area, and C_D is the drag coefficient.

The ability of trees to streamline reduces the cross-sectional area of the tree (A) and correspondingly the wind interception. Thus the drag for trees becomes more nearly linearly proportional to V. Alternatively one can say that the drag coefficient decreases as V increases.

The drag coefficient is a ratio expressing the force exerted on an object in relation to the maximum force which would occur if all the air flow was intercepted. Experiments have been carried out in wind tunnels to derive values of drag coefficients for trees: Mayhead (1973) gives values of 0.14 for western hemlock (*Tsuga heterophylla*), 0.29 for Scots pine and 0.36 for grand fir (*Abies grandis*). The variation was attributed to grand fir having dense foliage and being relatively rigid, with western hemlock, at the other extreme having very supple fine branches and short needles. As far as we are aware, similar experiments have not been carried out with broadleaved trees. There is a need for such tests, which, ideally, should be carried out on trees with and without leaves.

Tree failure

A tree may fail by uprooting, stem breakage or branch breakage. Coniferous and broadleaved trees suffer all these types of failure. Branch and stem breakage indicate either that stress in wood above ground is greater than stress in root wood or that the strength of root wood (minus the effect of the soil mass) is greater than that of aerial wood. Shed branches are likely to be carried in strong wind and can cause localised damage and sometimes injure or kill. Entire trees, however, can cause major structural damage, injury and death, and significant disruption of infrastructure.

Horse chestnut (*Aesculus hippocastanum*) appears to shed its branches as a means of survival, doing so regularly in high winds, and it is seldom seen uprooted (Cutler *et al.*, 1989).

Investigations into stress distribution along the trunk indicate that trees tend to adopt the form of a taper which creates an equal resistance to bending along its entire length (Leiser and Kemper, 1973).

Static and dynamic forces

For a tree to be uprooted, external forces must overcome the restoring moment of the tree. Experiments have been carried out using hand winches and pulley systems in which static loads have been applied to the tree until the failure point has been reached. In Sitka spruce (*Picea sitchensis*) forces of 10–52 kNm have been required to uproot 20-m high trees on brown earths and peaty gleys (Fraser and Gardiner, 1967; Coutts 1986) and of 3-14 kNm for 10-m high trees (Blackburn *et al.*, 1988).

Assuming that the resultant of the wind forces acts at two-thirds of the tree height, the moment of the drag force on the tree canopy can be calculated and, consequently, the corresponding wind velocities required to cause uprooting. In practice, however, the wind speed required to uproot trees is lower than that predicted by static loading (Cannell and Coutts, 1988).

Experiments with conifers have shown that considerations of mean wind speed alone cannot explain all the wind throw observed (Blackburn *et al.*, 1988). Wind is gusty and turbulent. Depending on the energy and frequency of the gusts, trees may begin a rocking motion. Because of this dynamic motion it appears that lower wind speeds than might be expected will uproot trees, particularly if the size and frequency of the gusts coincide with the natural sway period of the trees, thereby inducing resonant motion.

Sway periods of 1.5–5 s have been recorded for various conifers (Sugden, 1962; Holbo *et al.*, 1980; Milne, 1988). These sway periods depend on the physical parameters of the tree: height, centre of gravity, trunk diameter, trunk taper, wood density and elasticity as well as many others. The relative importance of each of these parameters is not clearly understood. Further complications are added by trees not being uniform throughout their structure but having inherent weaknesses at branching points and forks. The crown also undergoes rotational motions and the branches may appear to move independently in different directions and at different oscillation frequencies. This is likely to introduce damping effects, but their extent is so far unknown.

Roots and soil

Roots are influenced by the presence or absence of surrounding trees, soil type, soil profile, water availability and any soil disturbance or compaction. From a structural aspect, the smallest and most distal roots are of least importance. In our experience, trees uprooted in high winds usually have root plates < 5 m in diameter (even for the tallest trees), with roots either snapped at the edge or protruding from the soil of the root plate and snapped at smaller diameters.

As a tree falls, it does so about a fulcrum (centre point of rotation), on the leeward side. The further this fulcrum is from the centre of the tree, the more stable the tree. The fulcrum point is dependent on the number of main lateral roots radiating from the trunk, the angles between them and their thickness and taper (Coutts, 1983). Generally, there are a few thick roots which resist bending close to the trunk, but their stabilising effect is reduced by the wide angles between them. This is summarised in the following equation:

$$L = R \cos\left(\frac{b}{2}\right)$$

where L is fulcrum length, R is distance down the root to point of maximum bending and b is the angle between roots (Coutts, 1983).

L is very sensitive to the number of roots when there are fewer than six evenly spaced large laterals, because of the effect on b. The few cases studied showed that the fulcrum for a broadleaved tree is close to or inside the trunk.

On the windward side, the roots are subject to tension. If the tree falls they will be broken or pulled through the soil. Breakage depends on the ultimate tensile force (F_{ult}) which can be withstood by the roots. Lewis (1986) found that for sitka spruce

$$F_{ult} = 0.94 + 14.8\,A$$

where A is the cross-sectional area of root (mm^2) and F_{ult} is measured in N.

This indicates rather less strength than that found by Turmanina (1965), who obtained mean tensile strength values of 36 N mm^{-2} for species of poplar (*Populus*), birch (*Betula*) and oak (*Quercus*).

The contribution to stability by smaller roots is undetermined, as the area of soil failure beneath the roots will determine the mass of soil acting against the uprooting moment. The soil mass held together by small roots may well, therefore, play an important role in tree stability. Once soil cohesion is broken beneath a tree, a rocking motion induced by the wind may break the roots progressively, gradually lessening the tree's stability.

Whether a root breaks or is pulled through the soil depends on its tensile strength and on soil root adhesion. This, in turn, will vary seasonally with moisture content, soil type and the degree of branching within the root system. Root branching differs with the tree species, but in general roots developed in fine-textured soils are shorter and more branched than those in coarser, well-aerated soils, where roots tend to be longer, thinner and straighter (Ruark *et al.*, 1982).

Coutts (1986) produced a diagram (Figure 13.3) summarising the relative importance of the rooting components in sitka spruce on a peaty gley soil. The same principles of stability apply to broadleaved trees, although the values and relative importance of each component may be different.

Although lateral roots provide most stability, rooting depth is also important: an increase confers better tree stability (Fraser, 1962). The 'tap' root, although the first root produced by the tree, seems to add little to tree stability once lateral roots are established. In the trees we have observed the 'tap' root appears to have died when only 4–5 cm in diameter, if evident at all.

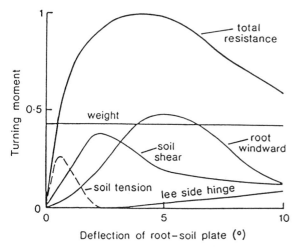

Figure 13.3 *Representation of components of total turning moment during uprooting of trees (from Coutts, 1986).*

Experimental work is being carried out to monitor the movement of the trunk, crown and root plate with simultaneous wind speed and wind direction readings. It is intended to produce power spectra of wind energy and corresponding tree displacement and also to calculate the tree sway periods of a selection of broadleaved trees. With the use of video and photographic techniques and static winching of trees it is hoped to gain an estimate of the drag coefficients for trees with and without leaves. Selected trees may then be taken beyond failure to attain values for the static overturning forces.

Much of this work depends on climatic conditions, control experiments giving way to measurements of extreme behaviour at opportune moments. Wind tunnel measurements may be a source of data, but would require a large wind tunnel, and gusting cannot be realistically simulated. It would not be possible to observe the entire tree, as, even if the roots were retained, the soil–root interaction would be lost. Observation of root plates also poses a problem as, generally, only those which have failed, (i.e. been uprooted by the wind) can be studied.

Conclusions

The complex nature of the factors involved in tree stability have been highlighted. It has been shown that uprooting is a dynamic problem involving the interplay of wind, tree oscillation, rooting regime and ground and topological conditions. Although a simple solution to the prediction of tree stability is not envisaged, it is hoped that results gained from further statistical analysis and practical measurements and observations will help to elucidate the problem.

REFERENCES

BLACKBURN, P., PETTY, J.A. and MILLER, K.F. (1988). An assessment of the static and dynamic factors involved in windthrow. *Forestry*, **61**, 29–43.

CANNELL, M.G.R. and COUTTS, M.P. (1988). Growing in the wind, *New Scientist* **21**, (January) 42–46.

CANNELL, M. and MORGAN, J. (1987). Young's modulus of sections of living branches and tree trunks. *Tree Physiology* **3**, 355–364.

COUTTS, M.P. (1983). Root architecture and tree stability. *Plant and Soil* **71**, 171–188.

COUTTS, M.P. (1986). Components of tree stability in Sitka spruce on peaty gley soil. *Forestry* **59**, 173–197.

CUTLER, D.F., GASSON, P.E. and FARMER, M.C. (1989). The wind blown tree root survey. *Arboricultural Journal* **13**, 219–242.

FRASER, A.I. (1962). The soil and roots as factors on tree stability. *Forestry* **35**, 117–127.

FRASER, A.I. and GARDINER, J.B.H. (1967). *Rooting and stability in Sitka spruce*. Forestry Commission Bulletin 40, 28. HMSO, London.

GIBBS, J.N. (1991). Survey of parkland trees in southern England after the gale of October 1987. In *Research for practical arboriculture*, ed. S.J. Hodge, Forestry Commission Bulletin 97. HMSO, London.

HOLBO, H.R., CORBETT, T.C. and HORTON, P.J. (1980). Aerodynamic behaviour of selected Douglas fir. *Agricultural Meteorology* **21**, 88–91.

LEISER, A.T. and KEMPER, J.D. (1973). Analysis of stress distribution in the sapling tree trunk. *Journal of the American Society for Horticultural Science* **98**, 164–170.

LEWIS, G.J. (1986). Root strength in relation to

windblow. *Forestry Commission Report on Forest Research 1986,* 68. HMSO, London.

MAYHEAD, G.J. (1973). Some drag coefficients for British forest trees derived from wind tunnel studies. *Agricultural Meteorology* **12,** 123–130.

MILNE, R. (1988). The dynamics of swaying trees. *New Scientist* **21,** (January), 46.

ROBERTSON, A. (1987). The use of trees to study wind. *Arboricultural Journal* **11,** 127–143.

RUARK, G.A., MADER, D.L. and TATTAR, T.A. (1982). The influence of soil compaction and aeration on the root growth and vigour of trees. A literature review – Part 1. *Arboricultural Journal* **6,** 251–265.

SUGDEN, M.J. (1962). Tree sway period – a possible new parameter for crown classification and stand competition. *Forestry Chronicle,* (September), 336–344.

TURMANINA, V. (1965). The strength of tree roots. *Bjulleten Moskovskogo Obscestva Ispytatelei Prirody (Otdel Biologiceskij)* **70,** 36–45. Cited by O'Loughlin, C. and Watson, A. (1979). Root-wood strength deterioration in radiata pine after clearfelling. *New Zealand Journal of Forestry Science* **9,** 284–293.

Discussion

T.H.R. Hall (Oxford University Parks)
Are you aware of the work done in Germany, particularly that of Dr Wessolly in Stuttgart on strength and stability of trees and of Claus Mattheck on the mechanics of trees? Did anyone actually watch a tree fall over and study how it moved before and during collapse?

H.J. Bell
Yes, I have heard of the German work and I would be very pleased to have any observations from people who watched the behaviour of trees during the storms.

C. Wm. Jorgensen (Consultant)
In observing trees, I have noted that in mixed woods mostly Scots pine and larch were affected, larch with a quarter-girth of 8–14 inches. Larch has no leaves in January and is a conifer (said to be less affected). This is possibly because it is grown in better soils.

H.J. Bell
Yes. We should look at soils and whether the wind was channelled.

C. Yarrow (Chris Yarrow & Associates)
I manage woodland that was badly damaged in October 1987, particularly some 30-year-old Scots pine yield class 14. In the strong north-east winds in January 1990 a lot of the remaining trees were blown over. Examination showed that the wood on the down-wind side appeared to be dead. Have you seen this effect on the formation of compression wood?

H.J. Bell
No, but this is something I would like to look at.

R.R. Finch (Roy Finch Tree Care)
At Westonbirt Arboretum there were many cypresses that had stood up. This species has a very fibrous root system. Larch has fewer, larger, thick roots and seems to sway with these roots as a fulcrum.

A. Coker (Department of Transport)
M.P. Coutts has done some work on this aspect. Is it possible to predict the point of failure from the design of the root system?

H.J. Bell
Yes, I think it is. Some work has been done on this.

A. Coker (Department of Transport)
Will you be able to give hard figures after this study?

H.J. Bell
I like to think so, but one would also have to consider the various crown structures.

J. McCullen (Office of Public Works, Dublin)
Has any cognizance been taken of the land-use around trees, animal compaction, for example?

H.J. Bell
I have noted ground cover, and any signs of compaction but this has not been quantified.

J.N. Gibbs
Not in the Forestry Commission study, where all the trees were in parkland.

R.P. Denton (Robert Denton Associates Ltd)
I am very impressed by the work that has been done although there seems to much con-

troversy about the causes and conflicting evidence about defects and susceptibility. Not enough account has been taken of wind factors. Trees oscillate in wind. Are we paying enough attention to the problems of turbulence and vortices on trees? These have a more significant role than defects or directional winds.

C.J. Baker (Nottingham University)

I think we can quantify the effect of turbulence on a tree, but cannot determine the transfer of wind force or energy to the tree.

C.P. Quine (Forestry Commission)

The effect of wind forces on buildings is fairly well understood. Trees are different as they are more dynamic structures.

Paper 14

Tree root survey by the Royal Botanic Gardens Kew and the Countryside Commission Task Force Trees

D.F. Cutler, *Jodrell Laboratory, Royal Botanic Gardens, Kew, Richmond, Surrey, TW9 3AB, U.K.*

Summary
Information from the survey of roots of windblown trees by Kew with Task Force Trees carried out after the gales of October 1987 is briefly discussed and suggestions for future research are outlined.

Survey
In general, research on roots has to be opportunistic. It is expensive and time consuming to excavate root systems, though this has been done for some fruit trees and plantation trees, and a few others. Most knowledge on roots has come from (i) the long-term survey set up at Kew, in which roots found in excavations at the foundations of damaged buildings were identified and then traced to the nearest tree of the particular species, giving some idea of root lengths, and (ii) the more recent Kew and Task Force Trees survey after the storm in October 1987.

We now have data on the root types of about 130 tree species and varieties up to the edge of the root plates that were exposed when the trees fell. These are illustrated and described, together with information on tree stability in a range of soils, by Cutler *et al.* (1990) and Gasson and Cutler (1990). The most frequent form of root system was one with lateral roots, with or without droppers. These were observed in over 81% of the trees surveyed. Sloping roots represented only 12.4% and trees with tap roots only 2.4% of the total.

The deepest roots exposed at the root plates were within 0.5 m of the surface in 15.2% of trees studied, and only 3.5% had roots deeper than 2.0 m. The remaining 81.3% of trees had their deepest roots between these depths. Of considerable interest is the narrow diameter of roots at 1 m from the trunk; 76.8% of the trees recorded had no roots over 10 cm in diameter. These data are important in relation to setting adequate building distances from trees, ensuring the survival and safety of the trees.

Future research
We now have reasonable knowledge of the nature of root systems close to trees, and some distance from them, but little is documented of what goes on in between. Using methods similar to those in 'rescue archaeology', it may be possible to set up a unit to record data on root systems exposed, for example, as a result of road building or open cast working.

The tension between builders and those caring for trees would be eased if we had data on the fate of trees close to which buildings had been erected. This would involve the design of a form to be filled in over a period of years, possibly by the department that gave advice on the 'safe' working distance. The completed forms could be sent to a central body for analysis. Details of the environment and health of the tree would be required, taking care to ensure that any factors that might affect the health of the tree were recorded. We know something of the roots; let us find out something sound about the result of damage to the roots.

It took some time to produce the questionnaires after the 1987 storm. We have learned something about form design as well as about roots! It would be wise to produce a new form in case an opportunity should arise to study roots from trees in other parts of the country and on other soil types. In particular, we missed a lot of information about urban trees in 1987. If those who might be expected to assist in a future survey had a form in their files ready for photocopying, they would be ready to collect the required data should the opportunity arise.

REFERENCES
CUTLER, D.F., GASSON, P.E. and FARMER, M.C. (1990). The wind blown tree survey: analysis of results. *Arboricultural Journal* **14**, 265–286.

GASSON, P.E. and CUTLER, D.F. (1990). Tree root plate morphology. *Arboricultural Journal* **14**, 193–264.

Discussion

T.P. Marsh (Southampton City Council)
In a storm crisis local-authority and commercial staff are too busy to complete forms; the priority must be clearance to allow life to continue. However, many root plates remain. Is the Kew survey to be continued?

D.F. Cutler
The intention is to publish the results accumulated and then take stock.

A. D. Bradshaw (Liverpool University)
The need is to match information on trees not blown over to assess the full extent of rooting by digging up standing trees.

D.F. Cutler
'Form B' asked for information about the proportion of trees still standing. Hydraulic washing-out of roots, as done for fruit and forestry trees, would be very expensive.

A. Lyon (London Borough of Tower Hamlets)
Many trees have apparently survived unscathed by the winds. How might trees with damaged roots be identified?

D.F. Cutler
Some trees died back after the storm; this may be progressive, but some trees have shown signs of recovery.

P.G. Biddle (Consultant, Arboricultural Association)
There is considerable valuable information from examination of trenches dug through root systems. The distance of root measurement from the tree stem reported by Dr Cutler is considerably shorter than recommended in British Standard (BS) 5837 (revision) for protective fences around trees. Stability of protected trees on development sites should not be impaired if the revised BS figures are used for protective fences.

J.A. Dolwin (Dolwin and Gray)
There is concern in the industry that a greater area of protection is needed if severe root severance is to be avoided on building sites.

D.F. Cutler
The data from the survey have been made available to the British Standards Institution.

S.A. Neustein (Forestry Commission)
Root examination in forest trees has lead to descriptions of root system types. There appears to be as wide a difference within as between species. The soil must, therefore, be considered to have a major influence on root development. The Forestry Industry has experienced several gales and a contingency plan has been drawn up for woodlands. The final point of the plan is to discourage the headlong rush into surveys. Wind is not regular, it is like the water in a shallow, bubbling brook crossing stones. If a subject is important, spend time making things happen. Sacrifice trees if necessary so that their reaction can be studied.

D.F. Cutler
The survey we undertook considered the soil, and the report will include reference to soil type. The extreme variability of the wind is acknowledged.

C. Yarrow (Chris Yarrow and Associates)
Sweet chestnut coppice was blown over in 1987. For stability, the root plates have been pushed back. What is the stability of the future coppice rotations likely to be?

D.F. Cutler
There is no information.

C.G. Bashford (Colin Bashford Associates)
Many trees appeared to have been rocked by the wind creating a socket around the stem

base. How badly have the roots been damaged?

D.F. Cutler

Some roots may have broken. Waterlogging in the socket and freezing of the water may cause further damage.

T. Walsh (Birmingham City Council)

The recommendations of the National Joint Utilities Group (NJUG) require that service trenches are 1.2 m from the kerbline. As a result, trees are likely to be repeatedly damaged.

D.F. Cutler

As roots tend to grow under pavements, trenching will damage roots unless hand digging can be undertaken.

P.G. Biddle (Consultant)

The NJUG has agreed to discuss the interaction of trees and services with a view to developing comprehensive guidelines.

C.L.A. Davis (Department of the Environment)

Why, in many instances, does one massive root develop in one direction on apparently uniform soil?

D.F. Cutler

There is no clear information. Water and nutrient supply probably have the greatest effect. Topography may have some influence.

Tree health

Paper 15

Air pollution and tree health in relation to arboriculture

J.E.G. Good, *Institute of Terrestrial Ecology, Bangor Research Unit, University College of North Wales, Deiniol Road, Bangor, Gwynedd, LL75 2UP, U.K.*

Summary

Increasing concern over air pollution as a possible cause of forest decline has led to an intensification of research into the extent and characteristics of air pollution in Europe, and the effect of these pullutants on trees. The chief air pollutants affecting trees and their geographical distribution, the mechanisms by which they are brought into contact with trees and the ways in which trees respond to them are identified. Special factors determining the influence of air pollution on urban trees are discussed. The trend of reducing sulphur dioxide emissions and the probable future requirement for catalytic converters to reduce hydrocarbon emissions from car exhausts indicate sustained improvement in the quality of the urban environment to the continuing benefit of urban trees and the choice of species that can be used by arboriculturists.

Introduction

It was well known in the nineteenth century that trees growing in towns and cities are damaged by air pollution, that conifers are generally more susceptible to damage than broadleaved trees and that some species of each are more tolerant of pollution than others (Brown and Nisbet, 1894). Practical problems in growing trees, especially conifers, in urban areas continued to be noted, which was the reason for the removal of the British National Pinetum from urban Kew to rural Bedgebury in 1924. By the late 1950s, air quality in towns and cities was improving as a result of the passing by Parliament of a series of Clean Air Acts (1956) limiting urban air pollution. The opportunities for increased and more varied tree planting presented by these improvements were noted in a British Government publication of the day (Ministry of Housing and Local Government, 1958).

As the levels of air pollution in British cities continued to decline in the 1960s, reports began to appear from Scandinavia of acidification of lakes and rivers and of damage to tree health. In the early 1970s the Swedish Government presented a paper at the United Nations Stockholm Conference on the Environment which drew attention to the long-range transport of pollutants and the possible effects of acid deposition on forest growth. The results presented in this and other contemporary papers, although conflicting, drew attention to 'acid rain' as an environmental problem and led to a major programme of research in Europe.

During the mid-1970s, concern grew in West Germany over the widespread occurrence of unexplained disease symptoms in silver fir (*Abies alba*) (Fink and Braun, 1978). This 'decline' in silver fir, which was not unprecedented, was followed by a much more widespread, but apparently different, decline in other species, notably Norway spruce (*Picea abies*), which is the dominant forest tree species in West Germany. This problem was taken very seriously by 1980, and in 1984 the 'Research Advisory Council of the German Government on Forest Damage/Airborne Pollutants' concluded that this was a new type of forest decline and stated that about 20% of the total forest area of the German Federal Republic was moderately or severely affected (Forschungsbeirat Waldschäden, 1984). Air pollution was proposed as the regional factor causing the decline, although opinions differed as to the pollutants concerned and the mechanism of damage (Ulrich *et al.*, 1980; Rehfuess, 1981; Blank, 1985).

During the 1980s there was growing apprehension over an apparently similar type of forest decline in red spruce (*Picea rubens*) growing in the Appalachian Mountains of eastern North America (Johnson, 1987). In this case too, it was proposed that air pollutants were the cause (Johnson and Siccama, 1983), although it was considered likely by some workers that climatic stresses were also involved (Hamburg and Cogbill, 1988).

In both instances, the decline in tree health was greater at higher altitudes (McLaughlin, 1985; Ammer *et al.*, 1988). Observations since 1982 have quantified the scale and geographical extent of the decline, and in both continents clear evidence of decline is found at sites separated by 400 km or more (Blank, 1985; Johnson, 1987).

Awareness of these forest decline problems in mainland Europe and North America caused growing concern in Britain in the early 1980s that air pollution might be affecting the condition of trees here. As a result, the Forestry Commission initiated a forest health survey in 1984 (Binns *et al.*, 1985), which has been gradually extended annually (Innes and Boswell, 1989). A second project, referred to as the European Community Forest Health Inventory, was started in 1987 as a direct result of European Community legislation (Commission Regulation (EC) No. 1696/87). Unlike the earlier survey, it enables the condition of trees in Britain to be compared with those in other EC countries. The results of these inventories are described and their significance is discussed elsewhere in this volume (Innes, 1991).

The aims of this paper are to:

i. identify the main air pollutants affecting trees, the sources from which they come and the mechanisms by which they are brought into contact with trees;
ii. describe the effects of location on the composition and deposition of air pollutants;
iii. discuss the ways in which trees respond to the air pollutants to which they are exposed;
iv. identify any special factors affecting urban trees and their responses to air pollution; and
v. determine whether air pollution is of increasing or decreasing importance for arboriculture in Britain and abroad.

Air pollutants affecting trees

The term 'acid rain' is now widely used as an all-embracing synonym for atmospheric pollution, often with little or no indication that acid rain comprises a mixture of substances, varies greatly in compositon with place and time, and is only one of several important forms of air pollution. This lax use of teminology has led to widespread misunderstanding among foresters

Table 15.1 Major gaseous and particulate pollutants which may occur in polluted air (after Last, 1989).

	Gaseous pollutants	Particulate pollutants
Primary	Sulphur dioxide (SO_2) Nitric oxide (NO) } collectively known as Nitrogen dioxide (NO_2) } oxides of nitrogen (NO_X) Hydrocarbons Ammonia Carbon dioxide (CO_2)	Fuel ash Metal particles
Secondary	Nitrogen dioxide formed by oxidation of nitric oxide. Ozone (O_3) and other gases formed by the action of sunlight in mixtures of NO_X and hydrocarbons. Nitric acid (HNO_3) formed by oxidation of NO_X.	Sulphuric acid (H_3SO_4) and nitric acid (HNO_3) formed by the oxidation of SO_2 and NO_X, respectively. Reaction products (($NH_4)_2SO_4$, NH_4NO_3, etc.) of sulphuric acid and nitric acid with other atmospheric constituents, notably ammonia (NH_3).

and arborists about the threat posed to tree health by air pollution. In this discussion of the air pollutants affecting trees, the term 'acid rain' will, therefore, generally be avoided, referring instead to the individual pollutants and the mixtures in which they may occur in the atmosphere and the means by which they may come into contact with trees.

Air pollutants, which may arise from natural sources (e.g. biological decay processes) as well as from industrial activities such as power generation from fossil fuels and smelting of mineral ores, are *primary* (produced directly) or *secondary* (created in the atmosphere by chemical processes involving primary pollutants). Each of these groups includes gases and particles (Table 15.1).

Of these pollutants, SO_2, NO_X and O_3 are most frequently implicated in damage to trees. Sulphur dioxide is a product of combustion of fossil fuels, especially coal, and of the smelting of sulphide metal ores. The amounts produced in the United Kingdom from 1860 to 1987 are shown in Figure 15.1. Emissions in Britain, as in most other major industrialised countries, have declined steeply in recent decades, notably because of the replacement of house coal with cleaner 'smokeless' fuel for domestic purposes and the conversion of many domestic and industrial users from solid fuel to other forms of energy which produce less SO_2.

There is no comparable long-term historical information on production of NO_X because this has only been regarded as a serious pollutant and been monitored since the 1960s. However, over 5 years in the early 1980s, the output from fossil-powered power stations and industry declined, while emissions in vehicle exhausts increased substantially (Table 15.2), a trend which is likely to continue (Department of the Environment, 1987).

Ozone occurs naturally in the lower atmosphere because of downward transfer from its regions of origin in the stratosphere. The natural ozone is supplemented by ozone produced from NO_X and hydrocarbons by photochemical generation in the boundary layer between the upper and lower atmosphere. Fossil-fuelled power stations and vehicles emit large amounts of NO_X but vehicles, unlike power stations,

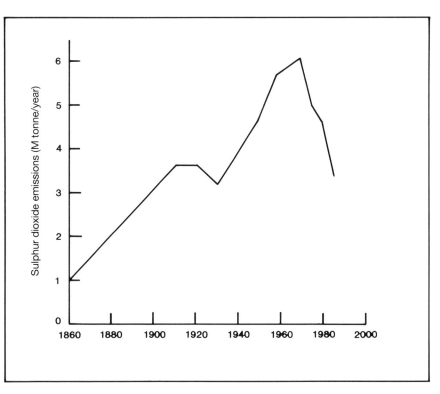

Figure 15.1 *Estimated United Kingdom sulphur dioxide emissions from 1860 to 1987 (after Chester, 1987).*

Table 15.2 Estimated United Kingdom annual emissions of oxides of nitrogen (t $NO_2 \times 10^3$) by source (from Department of the Environment, 1987).

Source of nitrogen oxides	1980	1981	1982	1983	1984	1985
Domestic	52	52	51	50	48	54
Commercial/public service	44	44	44	43	44	46
Power stations	851	819	767	763	621	738
Refineries	42	37	38	36	37	36
Agriculture (fuel)	4	4	3	3	3	3
Other industry	215	203	193	185	173	182
Rail transport	41	39	35	38	35	34
Road transport: petrol exhaust	477	468	494	503	513	517
diesel	177	166	172	185	203	216
Incineration/agricultural burning	12	12	12	12	12	12
All sources	1915	1844	1809	1818	1689	1838

also emit large amounts of hydrocarbons (Table 15.3). Thus, O_3 can be produced from vehicle exhaust emissions alone, but appreciable quantities can be produced in power station plumes only if the necessary hydrocarbons enter the plumes from other sources (Crane and Cocks, 1987). The importance of local conditions in ozone generation has been highlighted in the Los Angeles region of California, where damage to agricultural crops and trees has been shown to be due to ozone generated in photochemical smog. The British Isles has neither the climate nor the emissions required to produce photochemical smogs on the scale experienced in California. However, monitoring results have shown that ozone concentrations in summer are frequently above the natural background of 20–30 p.p.b. particularly in south-east England (Department of the Environment, 1987; Ashmore *et al.*, 1978) and that damage to plants may occur as a result.

The means by which gaseous and particulate pollutants can transfer to structures on the ground, including trees, are shown in Figure 15.2. Transfer may occur directly, by *dry deposition*, from atmosphere to trees and soils, or indirectly after incorporation into rain (so forming acid rain in the strict scientific sense), snow or

Table 15.3 Estimated United Kingdom annual emissions of oxides of hydrocarbons (t $\times 10^3$) (from Department of the Environment, 1987).

Source of hydrocarbons	1980	1981	1982	1983	1984	1985
Domestic	81	77	75	70	56	79
Commercial/public service	1	1	1	1	1	1
Power stations	14	14	13	12	10	12
Refineries (fuel use)	1	1	1	1	1	1
Other industry	4	4	4	4	3	4
Rail transport	10	10	9	10	9	9
Road transport: petrol exhaust	362	354	375	375	374	377
DERV exhaust	38	37	38	41	45	47
petrol evaporation	113	110	114	115	119	120
Petroleum refining/marketing	129	127	129	139	144	144
Gas leakage	349	349	367	366	384	419
Solvent evaporation	750	750	750	623	617	617
Industrial processes	–	–	–	121	121	121
Incineration/agricultural burning	38	38	38	38	38	38
Forests	71	71	71	71	71	71
All sources	1961	1943	1985	1987	1993	2060

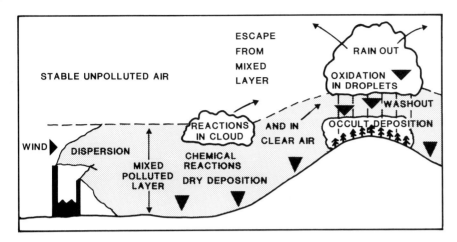

Figure 15.2 *Pathways by which atmospheric pollutants reach the ground (after Crane and Cocks, 1987).*

cloud/fog droplets. Rain and snow deposit under the influence of gravity, a process known as *wet deposition*, whereas cloud/fog droplets reach their targets by impaction or *occult deposition*, so named because the very small droplets involved are not trapped in standard rain gauges.

The chemical forms and spatial patterns of deposition depend strongly upon meteorological conditions, particularly on whether the plumes of polluted air encounter rain, and upon atmospheric concentrations of photochemical oxidants, such as ozone. It has been shown for SO_2 that dry deposition is easy to predict and generally declines in amount roughly in proportion to emission reductions (Crane and Cocks, 1987). Wet deposition, on the other hand, is complicated by non-linear chemical reactions involved in sulphate (SO_4) formation in rain and by the background amount of sulphate in the cloudwater which is not of pollutant origin.

Nitric oxide emissions are eventually oxidised, via nitrogen dioxide, to nitric acid (Table 15.1), which is rapidly deposited or removed in rain. However, the relationship between the amounts of nitric oxide in emissions and the amounts of nitric acid formed is complicated by the fact that oxides of nitrogen are involved in production of oxidants in the atmosphere.

Effects of location on deposition of pollutants

The mixture of pollutants in any particular body of air is influenced by:

the relative concentrations of natural, industrial, agricultural, vehicular and domestic sources;

interventions such as tallstack policies which alter the proportions of primary and secondary pollutants (e.g. higher chimneys increase the probability of larger volumes of SO_2 and oxides of nitrogen being converted to sulphuric and nitric acids);

the proximity of sources; and

altitude (particularly in relation to O_3 and occult deposition).

Using the analogy of different regional climates, and restricted by the limited data available, Last *et al.* (1986) have described areas in Western Europe with different mixtures of pollutants and identified them as having recognisable 'pollution climates'. The locations of three distinct pollution climates have so far been identified (Figure 15.3):

southern areas of the Federal Republic of Germany, including Baden-Wurttemberg and Bavaria, and most of Switzerland;

eastern Belgium, the Nordrhein–Westfalen region of Germany, western Czechoslovakia, and parts of central England; and

north-west Britain.

The dominant pollutants and/or deposition pathways in these climates appear to be:

Ozone, particularly at high altitudes, with significant inputs of S and N, primarily by wet deposition. Mists contribute significant

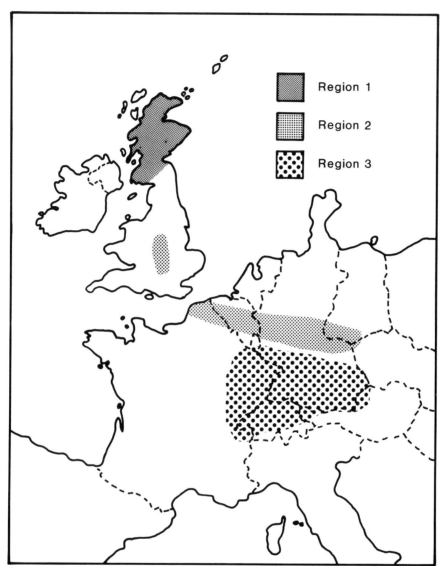

Figure 15.3 Map of western Europe showing the location of three regions with different pollution climates; see text (after Last et al., 1986).

amounts of sulphate, nitrate, and hydrogen ions.

SO_2 and NO_2 with a significant number of summer episodes of high ozone concentrations. Dry deposition dominates inputs of S, N and acidity.

Wet deposition of acidity, sulphate and nitrate and large concentrations of these ions in intercepted cloudwater.

It is clear from Figure 15.3 that large areas of Western Europe have not been characterised as having one of the three pollution climates and there must therefore be other such climates. Some of these are likely to depend upon local variations in climate. Jones and Bunce (1985) were able to identify the geographical location of 11 climate classes in western Europe and Scandinavia: four were northern and montane, four were temperate and three were Mediterranean classes.

Climate classes allow the possibility of predicting the return times (frequency per month, year, century, etc.) of extreme climatic events such as unseasonal frosts and droughts. These return times should give an indication of the expectation of climate-induced tree damage, while the climate classes *per se* may indicate where vegetation is more or less sensitive to pollutants, recognising that sensitivity is a function of a range of habitat variables (Last, 1989).

By combining information on the occurrence of pollution climates and climate classes, it should be possible to develop a clearer picture of how interactions between the two affect tree health. It is necessary to add information on soil types to the list because various studies (Ulrich, 1983; Zottl and Huttl, 1986; Rehfuess, 1988) indicate that the occurrence and severity of 'new type forest decline' in western Europe strongly reflects soil types. In general, trees growing on soils of low buffering capacity are said to be more severely damaged than those on soils of higher base status which can absorb more acid without themselves becoming acidified. In a recent review of the causes of one major type of spruce decline in Europe, foliar magnesium deficiency was shown to be responsible (Roberts *et al.*, 1989). Foliar Mg deficiency apparently results primarily from low rates of Mg uptake due to the low availability of Mg in some acid soils in central Europe. This low availability of Mg probably arises because losses of Mg into biomass over a single forest rotation and leaching of Mg from the soil by acid deposition greatly exceed exchangeable reserves in the soil plus atmospheric Mg inputs. It is interesting that this type of spruce decline has not been recorded in regions along the maritime seaboard of northern Europe (including the British Isles) where the rate of Mg deposition in precipitation is much greater than in central Europe.

Responses of tree to air pollution and causes of damage

The responses of trees to air pollution and how damage is caused are still far from fully understood. There are a number of reasons for the slow advance in knowledge in this area. First, as we have seen, trees in different locations experience different, often complex and varying pollution climates which, understandably, produce variable effects. These differences are confounded by interactions with climates and soils, giving many possible combinations of response (Last, 1989). Secondly, different tree species may respond in different ways to the same combination of adverse factors. Conifers are generally more sensitive than broadleaved trees and exhibit different symptoms of damage (Innes and Boswell, 1989). Within each of these major groups, species differ in susceptibility and sometimes in the symptoms they exhibit. There is also intraspecific variation in tolerance (Roberts, 1987). Thirdly, some of the symptoms of pollutant damage are similar to damage caused by other agencies, e.g. pests and pathogens, drought or nutrient deficiencies.

It is not practical to deal with all the many possible manifestations of pollutant damage and their causes when surveying tree health, or when researching the effects of air pollutants on trees. Therefore, symptoms of damage have been grouped into a number of recognisable distinct types. For example, five decline types for Norway spruce have been distinguished in the German Federal Republic (Forschungsbeirat Waldshäden, 1986). The symptoms range from needle-yellowing at higher altitudes of the German 'Mittelgebirge' (e.g. Black Forest) through needle-reddening of older stands in southern Germany (e.g. foothills of the Bavarian Alps) to thinning of tree crowns in coastal areas. Despite the variability of symptoms, all these types of decline are considered to be related and quite distinct from that observed in areas in Poland, East Germany and Czechoslovakia which have experienced direct damage by SO_2 from industrial sources for many decades. In these regions, the symptoms consist primarily of browning (necrosis) of the youngest shoots, which may be a direct effect of sulphur dioxide, or increased sensitivity to cold stress caused by exposure to SO_2 (Roberts, 1987).

Regardless of the initial symptoms, all serious air pollution damage results eventually in leaf loss, often with dieback of affected shoots. Crown density and leaf loss are, therefore, the most widely used indices for pollution damage. However, crown density does not necessarily reflect the health of a tree, nor does reduced crown density always indicate damage (Innes and Boswell, 1989). The distribution of foliage loss within the crown does, however, seem to give important indications of tree health. Thus, studies of Norway and Sitka spruce in Britain suggest that healthy trees retain needles for more than 7 years (Innes and Boswell, 1989), while studies in central Europe show that in Norway spruce less than 5–6 years of needle retention indicate that the trees are under stress (Heinsdorf *et al.*, 1988). As there is no difficulty in counting back 6 years, figures based on this index should be reliable.

The presence of dead branches within live crowns is also used as an index of tree condition. However, while this may be a satisfactory indicator for stands, where the proportions of trees with dead branches can be determined, it is less satisfactory for comparing the health of individual trees, as is necessary in arboriculture, because it is difficult to assess accurately the proportion of dead branches.

These general indicators of tree health may be applied to all species of conifers and broadleaves. In addition there are indices for particular species. For example, in beech (*Fagus sylvatica*) there is a condition known as leaf-rolling in which the normally flat leaves roll at their edges. Although the significance of leaf-rolling is still unknown, it is believed to be a symptom of ill health and is included in the assessment procedures recommended in the European manual for the assessment of forest condition (European Commission for Europe, 1988). Similarly, a high incidence of male flowers in Scots pine is believed to indicate stress and is used as an indicator of ill health.

There are probably many other characteristic responses to pollution not yet described, in particular the responses of individual tree species. Since arboriculture involves the use of many species and cultivars, the elucidation of such responses would be very useful to arborists, providing them with a guide (albeit imperfect) enabling pollution damage to be identified with more confidence than is possible now.

All of the indices described above refer to the condition of the tree, but it cannot be assumed that trees which show one or more of the symptoms of ill health have been adversely affected by air pollution. Most, perhaps all, of the symptoms can occur as a result of damage caused by natural agencies. There is thus a major problem in establishing cause and effect. Koch, a nineteenth century German pathologist, postulated three requirements for establishing cause and effect in disease (Koch, 1891):

> the suspected causal organism(s) must be constantly associated with the disease;
>
> the causal organism(s) must be isolated and identified; and
>
> when inoculated into healthy animals or plants, the causal organism(s) must reproduce the original disease.

The application of these postulates, which have been the touchstone for pathology since they were proposed, to 'pseudodiseases' such as forest decline is not easy, but it is argued by Last (1987) that it is necessary to make an attempt in order to impose discipline on investigations of causes and effects. The main difficulty arises from the fact that, as we have seen, forest decline is a variable 'disease' probably caused by a range of factors whose importance varies with place and time. In particular, cyclically changing patterns of weather, which are virtually impossible to mimic satisfactorily under controlled experimental conditions, seem to have an important role (Last, 1989). Nevertheless, it is possible to test the significance of particular pollutants to trees, either by excluding them from, or adding them to, a controlled environment, e.g. in an open-top chamber. Last (1987) emphasises the importance of ensuring that the controlled environment is related as closely as possible to the pollution climate associated with the geographical area in which the type of damage being investigated occurs. He also acknowledges the need to modify Koch's postulates by substituting 'pollutant(s)' for 'organism(s)' and rewriting the third postulate to read:

> pollutants deposited on plants (by wet and/or dry deposition) must help to produce the original disease.

By inserting the word 'help', Last accepts that pollutant(s) may not be the primary cause(s) of forest decline. This allows for the possibility that plants may be predisposed to damage by other adverse factors. Among these might be climatic stresses (e.g. frosts and droughts), soil factors (e.g. pH and nutrient status) and biotic factors (e.g. pests and diseases). Similarly, this modified postulate acknowledges that exposure to pollutants need not necessarily cause disease symptoms directly, but may do so by predisposing plants to damage by the adverse factors referred to above, or others.

As the various forms of forest decline have been investigated in more detail, attention being paid to the need to establish causes and effects, it has become clear that, in most instances where air pollution can be shown to be involved, the effects are either indirect or a combination of direct and indirect. Thus in Type 1 spruce decline the indirect factor of pollution-induced soil magnesium deficiency is causal (Roberts et al.,1989), while in decline of red spruce a combination of direct damage by acid mist and indirect damage resulting from reduced frost hardiness caused by exposure to acid mist seems to be responsible (Fowler et al., 1989; Leith et al., 1989). An example of trees apparently being climatically predisposed to injury by a specific pollutant is 'chlorotic decline' of ponderosa pine (Pinus ponderosa) in the San Bernardino Mountains near Los Angeles, California. This condition was associated with the production of fewer and smaller needles, yellow mottling of needles, the premature loss of all but the current season's needles, deterioration in systems of fibrous roots, decreased stem increments and ultimately death of some trees (Parmeter et al., 1962). Having eliminated pests, pathogens and graft-transmissible agents (viruses) as the primary causes of chlorotic decline, it was eventually shown in carefully controlled experiments that it could be induced in field conditions by ozone (Miller et al., 1963; Richards et al., 1968; Miller and McBride, 1975). However, the disease only came to the fore during a protracted period of drought from 1946 to 1960, which suggests that drought may have predisposed the trees to ozone damage.

The predisposition of trees to pest damage by exposure to air pollution is demonstrated by a study of the incidence of the aphid *Phyllaphis fagi* on beech seedlings grown in polluted and clean air. Fluckiger and Braun (1986) showed that air containing ambient concentrations of ozone favoured development of the aphid, compared with 'clean' air from which the ozone had been removed. It is thought that pollutants can encourage insect pests by increasing the concentrations of amino acids such as glutamine, arginine and proline in phloem exudates of plants.

It is clear that the causes of ill health (and, in some instances, improved condition) in trees associated with air pollution differ and may involve complex interactions between a number of factors in addition to the pollutants themselves. In some cases (e.g. Type 1 spruce decline) the causes are well understood, while in others investigations are still at an early stage.

Special factors affecting urban trees

Urban trees, and trees downwind of urban areas, are likely to be exposed to different pollution climates from trees growing in adjacent less polluted rural areas. The obvious difference is the generally greater total pollutant load but the relative proportions of different pollutants also differ. In urban areas in Britain, there has been a marked reduction in concentrations of SO_2 and soot since the Clean Air Acts of the 1950s and 1960s came into force. On the negative side, however, there has been a sustained increase in vehicle exhaust emissions which contain low concentrations of sulphur dioxide but large amounts of oxides of nitrogen and unburnt hydrocarbons giving rise to ozone (Department of the Environment, 1987). Ideal conditions for ozone generation from pollutants occur during periods of stable anticyclonic weather in summer. The peak rates of O_3 production do not necessarily occur over the area where the primary pollutants are produced. Studies of ozone generation in the London area, using aircraft with detection equipment and ground-level monitoring networks, showed that maximum concentrations are normally detected several hours downwind of the urban source of hydrocarbons and nitrogen oxides, different hydrocarbons reacting with nitrogen oxides at different rates (Department of the Environment, 1987).

Ozone 'episodes' are typical of the pollution climate found in much of central England (Last's second pollution climate), potentially phytotoxic concentrations occurring, on average, on about 40 days a year (Fowler *et al.*, in preparation). Thus, trees in many English towns and cities may be adversely affected by ozone from time to time. However, there is no proven evidence of visual damage, and more research is needed to establish whether this occurs and whether, in the absence of such damage, tree growth and susceptibility to other adverse environmental factors are affected by high ozone concentrations.

The surface structure of the built environment is much more variable than that of agricultural crop and forest canopies. High and low buildings are interspersed with hard surfaces at ground level and open spaces with or without vegetation, sometimes including trees. This structural variability affects the transfer of atmospheric pollutants to the ground, but in ways which are largely unknown (Fowler, 1984). Small-scale local variations in concentrations of pollutants ('pollution hot spots') are probably common in urban areas, reflecting local variations in the production of pollutants, influenced in some cases by non-random patterns of transfer. This should be borne in mind when considering possible cases of air pollution damage to urban trees; localised damage among an otherwise healthy tree resource should be expected.

Urban trees are subject to various stresses which may predispose them to damage by air pollutants. Soil surrounding the roots of trees growing in streets and near buildings is often of poor quality, containing large amounts of rubble and debris. Engineering specifications for streets, car parks and paved areas often require intensive soil compaction. Unfortunately, the places where trees are to be planted are frequently also compacted. Tightly compacted soils are poorly aerated, and gaseous exchange between soil air and atmosphere is restricted. This results in an unfavourable balance between oxygen and carbon dioxide that can inhibit tree root growth. Compacted soils are also poorly drained internally, water occupying air spaces and further restricting gaseous exchange. Poor soil aeration can also result from a rise in ground water tables. Water tables are frequently altered by construction activity.

Microclimatic factors having the greatest influence on tree growth are air temperature, humidity and wind (Grey and Deneke, 1978). In general, cities tend to be warmer in summer and winter than the surrounding countryside. Wind velocity is less and relative humidity is generally lower. Such generalities can be misleading, however, as cities are not single microclimates. Each location within an urban setting has its own microclimate, depending on the character and arrangement of the various townscape elements (man-made structures, land forms, trees and other organisms). Federer (1971) identified three broad classes of street-level microclimates: i. areas with extensive evaporative or transpiring surfaces, such as parks, wide streets with trees and the vicinities of rivers or lakes; ii. wide treeless streets, squares and car parks, open to the sky and very dry; and iii. narrow streets and courtyards surrounded by tall buildings.

Because of extensive transpiration, microclimate i tends to have lower temperatures and higher humidity in the summer. It is also cooler in the winter than other areas of the city. Wind velocities are higher because there are fewer physical barriers. These areas more nearly represent the climate of the surrounding countryside. From a climate standpoint, there would be little advantage or disadvantage for tree growth in these situations.

Microclimate ii has much lower humidity and higher temperatures than other areas. High temperatures tend to be extreme because of high radiation. Wind velocities are average to slightly lower than the surrounding countryside. Trees planted here are the most stressed and might be expected to be most predisposed to damage by air pollution.

Microclimate iii has cooler summer temperatures and much greater wind protection than other areas. Winter temperatures are somewhat warmer because of radiated heat from surrounding buildings. Trees planted here are likely to be less stressed than those planted in areas with microclimate ii.

Is air pollution a growing or declining threat to arboriculture?

The very wide range of tree species and cultivars grown successfully in the many rural and suburban arboreta in Britain suggests that air pollution has probably never been a serious threat to arboriculture in most parts of Britain. In many urban areas, however, air pollution (mostly SO_2 and soot) was so severe in the past as to limit the choice of reliable species to a few broadleaves known to be particularly tolerant of air pollution (see Introduction). Perhaps the best evidence that the atmosphere in British towns and cities is now more conducive to healthy tree growth is the much greater range of tree species and cultivars successfully cultivated in urban areas. I know of no hardy broadleaf, and few conifers, which could not now be planted with a good chance of success in all our towns and cities, including the most heavily polluted areas.

The major threat to tree health as SO_2 emissions decline is probably from pollutants, notably nitric oxide and hydrocarbons, in vehicle exhaust emissions. However, there is no evidence of damage being done to trees by these substances or others derived from them at present in Britain. The concentrations of hydrocarbons, and of ozone which is dependent for its production upon them, in vehicle exhaust emissions will, in any case, probably be reduced as catalytic converters become more widely used. These break down hydrocarbons to carbon dioxide and water. This means, of course, that the gain in terms of reduced hydrocarbon and ozone pollution is offset in general environmental improvement terms by the increased CO_2, which is a 'greenhouse' gas. This increased CO_2 is unlikely to have any direct adverse effect on tree growth and may, by stimulating photosynthesis, increase it. The wider global environmental significance of the extra CO_2 is not relevant to the present discussion.

The improved situation described above probably applies to most Western European and North American towns and cities but there are many countries in the world, including those in Eastern Europe, where pollution concentrations remain high or are continuing to rise. Arboriculture in such places will continue to be restricted, or in extreme cases prevented, by air pollution.

REFERENCES

AMMER, U., BURGIS, B., KOCH, B. and MARTIN, K. (1988). Untersuchungen uber den Zusammenhang zwischen Schadigungsgrad und Meereshohe im Rahmen des Schwerpunktoprogramms zur Erforschung der Wechselwirkungen von Klima und Wald schaden. *Forstwissenschaftliches Zentralblatt* **107**, 145–151.

ASHMORE, M.R., BELL, J.N.N. and REILY, C.L. (1978). A survey of ozone levels in the British Isles using indicator plants. *Nature* **276**, 813–815.

BINNS, W.O., REDFERN, D.B., RENNOLLS, K. and BETTS, A.J.A. (1985). *Forest health and air pollution: 1984 survey.* Forestry Commission Research and Development Paper 142. Forestry Commission, Edinburgh.

BLANK, L.W. (1985). A new type of forest decline in Germany. *Nature* **314**, 311–314.

BROWN, J. and NISBET, J. (1894). *The forester*. Blackwood, Edinburgh.

CHESTER, P.F. (1987). Acid rain – a prognosis. *CEGB Research* **20**, 62–64.

CRANE, A.J. and COCKS, A.T. (1987). The transport, transformation and deposition of airborne emissions from power stations. *CEGB Research* **20**, 3–15.

DEPARTMENT OF THE ENVIRONMENT (1987). Summary of ozone concentrations. In *Ozone in the United Kingdom, an interim report by the United Kingdom Photochemical Oxidants Review Group*, 29–61. London.

EUROPEAN COMMISSION FOR EUROPE (1988). *Manual on the methodologies and criteria for harmonized sampling, assessment, monitoring and analysis of the effects of air pollution on forests*. United Nations.

FEDERER, C.A. (1971). Effects of trees in modifying urban microclimate. In *Proceedings of symposium on the role of trees in the South's urban, environment*, 26–34. US Department of Agriculture Forest Service publication.

FINK, S. and BRAUN, H.J. (1978). Zur epidemischen Erkrankung der Weisstanne *Abies alba* Mill. I. Untersuchungen zur Symptomatik und Formulierung einer Virushypothese. *Allgemeine Forst- und Jadzeitung* **149**, 145–150.

FLUCKIGER, W. and BRAUN, S. (1986). Effect of air pollutants on insects and hostplant/insect relationships. In *How are the effects of air pollutants on agricultural crops influenced by the interaction with other limiting factors?* Workshop proceedings, March 1986. CEC, Riso National Laboratory.

FORSCHUNGSBEIRAT WALDSCHADEN (1984). *Forschungsbeirat Waldschäden / Luftverunreinigungen*. Zwischenbericht, December 1984.

FORSCHUNGSBEIRAT WALDSCHADEN (1986). *Forschungsbeirat Waldschäden / Luftverunreinigungen*. 2. Karlsruhe: Bericht.

FOWLER, D. (1984). Transfer to terrestrial surfaces. *Philosophical Transactions of the Royal Society of London, B* **305**, 281–297.

FOWLER, D., CAPE, J.N., JOST, D and BELKE, S. *The air pollution climate of non-nordic Europe*. (in preparation).

FOWLER, D., CAPE, J.N., DEANS, J.D., LEITH, I.D., MURRAY, M.B., SMITH, R.I., SHEPPARD, L.J. and UNSWORTH, M.H. (1989). Effects of acid mist on the frost hardiness of red spruce seedlings. *New Phytologist* **113**, 321–335.

GREY, G.W. and DENEKE, F.J. (1978). *Urban forestry*. John Wiley, New York.

HAMBURG, S.P. and COGBILL, C.U. (1988). Historical decline of red spruce populations and climatic warming. *Nature* **331**, 428–430.

HEINSDORF, D. KRAUSS, H.H. and HIOOELI, P. (1988). Ernahrungs- und bodenkundliche Untersuchungen in Fichtenbestanden des mittleren Thuringer Waldes unter Berucksichtigung der in den letzen Jahren aufgetretenen Unweltbelastungen. *Beitrage for Forstwirtschaft* **22**, 160–167.

INNES, J.L. (1991). Acid rain: tree health surveys. In *Research for practical arboriculture*, ed. S.J. Hodge, Forestry Commission Bulletin 97. HMSO, London.

INNES, J.L. and BOSWELL, R.C. (1989). *Monitoring of forest condition in the United Kingdom – 1988*. Forestry Commission Bulletin 88. HMSO, London.

JOHNSON, A.H. (1987). Deterioration of red spruce in the Northern Appalachian Mountains. In *Effects of atmospheric pollutants on forests, wetlands and agricultural systems*, eds T.C. Hutchison and K.M. Meema, 83–99. NATO, ADSI Series, 916. Springer-Verlag, Berlin.

JOHNSON, A.H. and SICCAMA, T.G. (1983). Acid deposition and forest decline. *Environmental Science and Technology* **17**, 294a–305a.

JONES, H.E. and BUNCE, R.G.H. (1985). A preliminary classification of the climate of Europe from temperature and precipitation records. *Journal of Environmental Management* **20**, 17–29.

KOCH, R. (1891). Uber bakteriologische Forschung. *Verhandlugendes 10th Internationaler Medizinischer Kongress 1890*, 35–47.

LAST, F.T. (1987). The nature and elucidation of causes of forest declines. In *Forest decline and reproduction: regional and global consequences*. Proceedings of a workshop held in Krakow, Poland, March 1987, eds L. Kairiukstis, S. Nilsson and A. Straszak, 15pp.

LAST, F.T. (1989). Experimental investigation of forest decline: the use of open-top chambers. In *Projekt Europaisches Forschungszentrum – Statuskolloquium*. March 1989. Karlsruhe: Kernforschungszentrum.

LAST, F.T., CAPE, J.N. and FOWLER, D. (1986). Acid rain or 'pollution climate'? *Span* **29**, 2–4.

LEITH, I.D., MURRAY, M.B., SHEPPARD, L.J., CAPE, J.N., DEANS, J.D., SMITH, R.I. and FOWLER, D. (1989). Visible foliar injury of red spruce seedlings subjected to simulated acid mist. *New Phytologist* **113**, 313–320.

McLAUGHLIN, S.B. (1985). Effects of air pollution on forests: a critical review. *Journal of the Air Pollution Control Association* **35**, 512–534.

MILLER, P.R. and McBRIDE, J.R. (1975). Effects of air pollutants on forests. In *Responses of plants to air pollution.* eds J.B. Mudd and T.T. Kozlowski. Academic Press, New York.

MILLER, P.R., PARMETER, J.R., TAYLOR, O.C. and CARDIFF, E.A. (1963). Ozone injury to the foliage of ponderosa pine. *Phytopathology* **53**, 1072–1076.

MINISTRY OF HOUSING AND LOCAL GOVERNMENT (1958). *Trees in town and city.* HMSO, London.

PARMETER, J.R., BEGA, R.V. and NEFF, T. (1962). A chlorotic decline of ponderosa pine in southern California. *Plant Disease Reporter* **46**, 269–273.

REHFUESS, K.E. (1981). Uber die Wirkungen der sauren Niedereschlage in Waldokosystemen. *Forstwissenschaftliches Centralblatt* **100**, 363–381.

REHFUESS, K.E. (1988). Ubersicht uber die bodenkundliche Forschung in Zusammnehang mit den Neuartigen Waldschaden. In *Projekt Europaisches Forschungszentrum-Statuskolloquium.* March 1988. Karlsruhe: Kernforschungszentrum.

RICHARDS, B.L., TAYLOR, O.C. and EDMUNDS, G.F. (1968). Ozone needle mottle of pines in southern California. *Journal of the Air Pollution Control Association* **18**, 73–77.

ROBERTS, T.M. (1987). Effects of air pollutants on agriculture and forestry. *CEGB Research* **20**, 39–52.

ROBERTS, T.M., SKEFFINGTON, R.A. and BLANK, L.W. (1989). Causes of type 1 spruce decline in Europe. *Forestry* **62**, 180–222.

ULRICH, B. (1983). Effects of accumulation of air pollutants in forest ecosystems. In *Acid deposition: a challenge for Europe*, eds H. Ottl and H. Stangl, 127–146. Proceedings of Symposium held by the Commission of the European Communities at Karlsruhe, September 1983. CEC, Karlsruhe.

ULRICH, B., MAYER, R. and KHANNA, P.K. (1980). Chemical changes due to acid precipitation in a loess-derived soil in Central Europe. *Soil Science* **30**, 193–199.

ZOTTL, H.W. and HUTTL, R.F. (1986). Nutrient supply and forest decline in south west Germany. *Water, Air and Soil Pollution* **31**, 449–462.

Discussion

R. Cromar (Arboricultural Company)
Can magnesium deficiency be made up from marine sources?

J.E.G. Good
There is no evidence from north-west Scotland. Remember that leaching will drive cations from the soil.

C. Yarrow (Chris Yarrow & Associates)
With regard to acid deposition on soil, is it too early for liming to halt forest decline?

J.E.G. Good
Magnesium sulphate has been shown to have the potential to reverse acidification in Europe.

J.L. Innes
Acid soils are sensitive to the effects of lime and hence liming may add to the problem.

Paper 16
Acid rain: tree health surveys

J.L. Innes, Forestry Commission, Alice Holt Lodge, Farnham, Surrey, GU10 4LH, U.K.

Summary

Surveys of tree health are being carried out in Great Britain and Europe. Those undertaken by the Forestry Commission have not indicated a widespread problem of forest decline in Britain. In some areas, problems have been identified, but these had already been recognised and explanations are available for most. No consistent link between air pollution and forest health has been identified, although various correlations are apparent. Generally, these have suggested that tree health is better in the more polluted parts of Britain than in less polluted areas but a causal relationship is not necessarily involved.

Introduction

Since 1984, the Foresty Commission has undertaken annual surveys of tree health in Great Britain. During the period, the programme has gradually developed as more information has been gathered. This has resulted in many changes to the original design and the study is now one of the most sophisticated of its kind. The surveys were established to assess the extent of forest decline in Great Britain and, if it was identified, to determine whether air pollution was present. Although these aims still exist, it is now apparent that forest decline may not be as widespread a problem as initially feared and the monitoring programme therefore has broader aims. These are essentially to investigate the annual changes in forest condition and to establish the reasons.

Current programme

The current programme arose out of the initial surveys and 1987 has been taken as the base year. This means that, in 1990, the results of three consecutive surveys are available. Within an international context, emphasis has been placed on two main indices: crown density and crown discoloration. These are assessed in the British survey, but a number of other indices have also been added (Table 16.1) so that each tree is comprehensively evaluated each year. In 1989, 7436 trees were assessed and there has recently been a decision to extend this to 9600 by 1991. For convenience, this programme is referred to throughout this paper as the main survey.

Five species are involved in the main survey: Sitka spruce, Norway spruce, Scots pine, oak and beech. The sample sizes are currently unequal, emphasis being placed on the three coniferous species, but the objective of increasing the sample size over the next 2 years is to increase the number of oak and beech trees assessed.

A further 1800 trees are examined as part of a European programme (the European survey), although the assessments of these are much less detailed than in the main survey. The European survey includes a wide range of species; it is based on a systematic sampling design, and the sampling frequency of each species (Table 16.2) is roughly proportional to its abundance in woodlands. Some of the samples are very small, and of little value in themselves, but, when they are aggregated with data from the rest of Europe, the information is of more value (cf. Commission of the European Communities, 1989).

Table 16.1 Characters assessed in a survey of tree health since 1987. (Ss, Sitka spruce; Ns, Norway spruce; Sp, Scots pine; Ok, oak; Be, beech).

Height (all). Measured once every 5 years.
Diameter at breast height (all). This, and all subsequent assessments, are made annually.
Dominance (all).
Crowding (all). Degree of canopy closure around the tree.
Crown form (all). The type of crown. Different forms for each species.
Crown density (all). Amount of foliage in the crown.
Defoliation type (Ss, Ns, Sp). Pattern of needle loss within the crown.
Branch density (Ss, Ns, Sp). The extent of influence of the growth rate of the tree on crown density.
Number of needle-years present (Ss, Ns, Sp).
Extent of shoot death in the crown (Ss, Ns, Sp).
Location of shoot death in the crown (Ss, Ns, Sp).
Dieback type (Ok, Be). Severity of dieback within the crown.
Dieback location (Ok, Be). Location within the crown.
Percentage of crown affected by dieback (Ok, Be). Contribution of dieback to crown density score.
Recent shoot growth (Ok, Be). A measure of the state of degeneration of the crown.
Leader condition (Ss, Ns, Sp).
Abundance of secondary shoots within the crown (Ss, Ns).
Location of secondary shoots on individual branches (Ss, Ns).
Number of stem epicormics (Ok).
Number of branch epicormics (Ok).
Extent of male flowering (Sp).
Fruiting extent (all).
Number of green leaves on ground under tree (Be).
Leaf size (Be).
Degree of leaf-rolling in crown (Be).
Frequency of rolled leaves in crown (Be).
Overall discoloration of foliage (all).
Extent of browning of current year's needles (Ss, Ns, Sp).
Extent of yellowing of current year's needles (Ss, Ns, Sp).
Type of yellowing of current year's needles (Ss, Ns, Sp). Various forms are recognised.
Extent of browning of older needles (Ss, Ns, Sp).
Extent of yellowing of older needles (Ss, Ns, Sp).
Type of yellowing of older needles (Ss, Ns, Sp).
Extent of browning of leaves (Ok, Be).
Extent of yellowing of leaves (Ok, Be).
Type of yellowing of leaves (Ok, Be).
Extent of mechanical damage to crown (all).
Type of mechanical damage to crown (all).
Extent of butt damage (all).
Type of butt damage (all).
Extent of stem damage (all).
Type of stem damage (all).
Extent of fungal damage to foliage (all).
Extent of insect damage to foliage (all).

In both surveys, trees are in areas of woodland of at least 0.5 ha, and isolated trees or trees growing in stands of <0.5 ha are excluded. At each sampling site, 24 trees, divided into four groups of six trees, are assessed. Each sub-plot is separated by at least 25 m. Full details of the sampling designs of the two surveys can be found in Innes and Boswell (1987).

Survey results

Results of both surveys have already been published (Innes and Boswell, 1987, 1988, 1989, 1990; Commission of the European Communities, 1989).

Crown density

By far the most frequently used measure of tree

Table 16.2 Numbers of trees sampled in the British portion of the United Nations/Economic Commission for Europe and Commission of European Communities surveys of forest health.

Species	<60 years old	>60 years old	Mixed-age stands	Total
Acer pseudoplatanus	15	15	32	62
Alnus glutinosa	39	1	1	42
Alnus viridis	0	4	0	4
Betula pendula	26	2	9	37
Betula pubescens	31	6	25	62
Carpinus betulus	20	0	4	24
Castanea sativa	30	0	4	34
Corylus avellana	0	0	8	8
Eucalyptus sp.	4	0	0	4
Fagus sylvatica	17	19	36	72
Fraxinus excelsior	30	31	18	79
Ilex aquifolium	0	0	6	6
Populus nigra	13	0	0	13
Populus tremula	1	0	0	1
Prunus avium	2	2	2	6
Quercus petraea	2	24	2	28
Quercus robur	31	55	50	136
Quercus rubra	2	0	0	2
Salix sp.	0	0	1	1
Sorbus aucuparia	5	4	1	10
Tilia cordata	0	2	0	2
Ulmus glabra	2	0	0	2
Ulmus minor	0	0	9	9
Other broadleaved trees	28	2	6	36
Abies alba	1	0	0	1
Larix decidua	28	0	0	28
Larix kaempferi	63	0	3	66
Picea abies	48	0	0	48
Picea sitchensis	507	0	7	514
Pinus contorta	126	0	0	126
Pinus nigra	31	0	0	31
Pinus radiata	0	0	6	0
Pinus sylvestris	210	0	26	236
Pseudotsuga menziesii	62	1	8	71
Thuya sp.	3	0	0	0
Tsuga sp.	3	0	0	0
All broadleaved trees	298	167	214	679
All conifers	1082	1	50	1133
Total	1380	168	264	1812

condition is crown density. This is assessed by estimating the amount of light passing through the crown. Reference photographs (Bosshard, 1986; Innes, 1990a) or block diagrams (Belanger and Anderson, 1988) are used, illustrating trees pictorially or schematically with reducing crown densities. Crown density does not necessarily reflect tree health (Westman and Lesinski, 1986; Mahrer, 1989), nor is there a consistent relationship between crown density and growth (Innes and Cook, 1989). For example, the densest trees may also be the slowest growing and trees with thin crowns may be so because they are growing rapidly (as is the case

Table 16.3 Percentage of trees in each crown density class for five species in Britain in 1987–1989.

		Crown density class*									
	Year	0	1	2	3	4	5	6	7	8	9
Sitka spruce	1987	14	23	25	19	12	5	2	0	0	0
$n = 1223$	1988	8	22	26	23	14	6	1	0	0	0
	1989	13	22	24	19	10	5	3	2	2	0
Norway spruce	1987	21	24	24	15	9	4	2	1	0	0
$n = 1512$	1988	21	27	24	16	7	4	1	0	0	0
	1989	27	28	22	14	6	2	1	0	0	0
Scots pine	1987	19	23	25	18	8	3	1	1	1	1
$n = 1459$	1988	8	23	32	20	9	4	1	1	1	1
	1989	13	31	27	15	6	3	2	1	1	1
Oak	1987	9	15	21	29	15	6	4	1	0	0
$n = 695$	1988	4	16	29	27	14	6	2	1	1	0
	1989	7	22	31	26	10	3	1	0	0	0
Beech	1987	8	22	26	28	12	3	1	0	0	0
$n = 672$	1988	9	24	33	22	9	2	1	0	0	0
	1989	19	31	29	16	4	1	0	0	0	0

*Proportion of light passing through the crown in 10% intervals, thus class 0 = 0–10%, class 1 = 10–20%, class 2 = 20–30%, etc. of light passing through; crown density decreases as number of class increases.

in many British trees).

The results from the main survey for 1987–1989 are presented in Table 16.3; only trees that were assessed in all 3 years are included. As the sample has changed over the last 3 years (because of loss and replacement of plots), the sample sizes do not reflect the total number of trees assessed in each year. Trees have been divided into density classes of 10%, class 0 representing trees allowing 0–10% of light to pass through the crown, class 1 being 10–20% and so on.

Crown density generally improved (i.e. was greater) in 1989 compared with 1987 or 1988. In beech, there has been a steady improvement since 1987; the other species do not show a consistent trend over the 3 years. The proportion of trees in density classes 6–9 (i.e. allowing >60% of light to pass through the crown) is small. Only Sitka spruce has shown a significant increase in this category and this can be directly attributed to heavy infestations by the green spruce aphid *Elatobium abietinum* in 1989, which resulted in severe defoliation of some trees.

It is difficult to judge from these data what constitutes the normal situation. There is no information on what the distribution of crown density classes would look like in the absence of air pollution. However, as all the trees sampled in the main survey were outside urban areas, they probably provide a better picture than those within towns (which not only are subjected to pollution but also have to withstand a variety of other stresses, such as root compaction and salting). The data for 1989 are presented as frequency histograms in Figure 16.1, using 5% reduction in density classes. As in Table 16.3, the higher the category, the lower the density of the crown. All the distributions are highly skewed. The general pattern for each species is similar despite their differing resistance to pollution. The exception is Sitka spruce, which, as already stated, suffered from severe insect defoliation in many locations in 1989.

Because of the skewed distributions, shown in Figure 16.1, it is difficult to determine confidence limits for the distributions, which might help to determine whether individual trees were

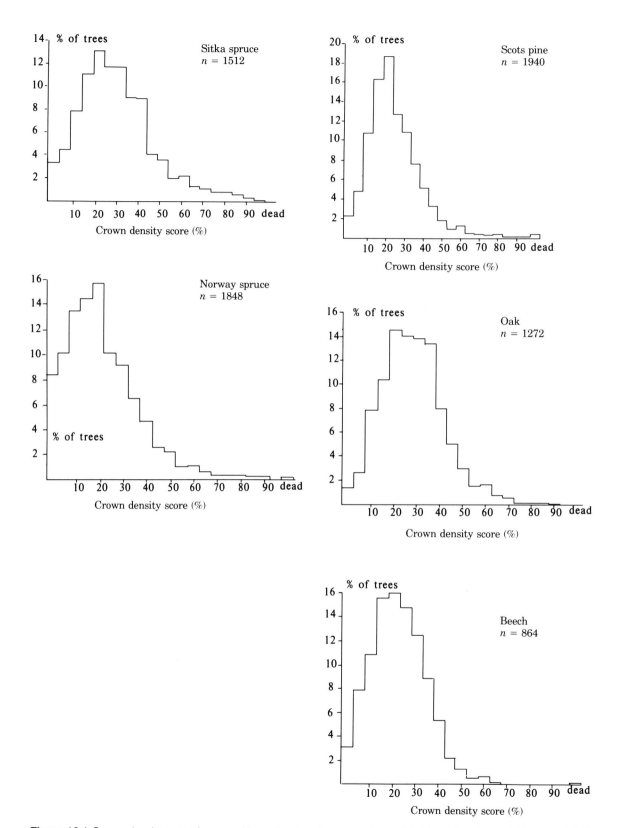

Figure 16.1 *Crown density score (assessed by estimating the percentage of light passing through the crown) in five tree species in Britain in 1989.*

Table 16.4 Confidence limits for mean site scores for crown density in five tree species surveyed in Britain in 1988 and 1989. See Table 16.3 for sample sizes.

	Year	−2 S.D.*	−1 S.D.	Mean	+1 S.D.	+2 S.D.
Sitka spruce	1988	10.94	19.45	27.96	36.47	44.98
	1989	4.33	16.74	29.16	41.57	53.98
Norway spruce	1988	–	13.13	22.86	32.59	42.32
	1989	–	9.96	20.45	30.94	41.43
Scots pine	1988	5.69	16.69	27.68	38.67	49.66
	1989	3.31	14.03	24.76	35.48	46.20
Oak	1988	14.63	23.11	31.59	40.08	48.56
	1989	12.43	20.31	28.20	36.09	43.97
Beech	1988	11.66	18.77	25.89	33.01	40.12
	1989	7.54	14.83	22.11	29.39	36.67

*Standard deviation

exceptional. However, confidence limits can be calculated for the mean site crown density scores. This has been done for both 1988 and 1989 (Table 16.4). The data have been calculated for one and two standard deviations (S.D.) about the mean. No data are available for the lower end of the Norway spruce range as the limit falls below the minimum measurement possible.

The figures are fairly stable for each of the species over the 2 years. The standard deviations for Norway spruce, Scots pine, oak and beech remained constant, but there was a significant increase in Sitka spruce. Generally, the two broadleaved species were less variable than the coniferous species despite oak having a higher mean score. Ideally, data should be gathered for several more years before the confidence limits are used in a predictive fashion, but they should provide an approximate guideline for assessing the current relative condition of stands. The figures are based on the average for 24 trees at a site; data for individual trees cannot be used.

Other indices

A variety of other indices was assessed (Table 16.1); many were introduced in 1989 and the results are still being analysed (Innes and Boswell, 1990).

Discoloration in the conifers and in oak was rare (Table 16.5), fewer than 20% of the trees being affected. Scots pine tended to show more discoloration than the other two conifers, but this may be related to the development of autumnal senescence of the older needles towards the end of each survey period. The highest discoloration scores were for beech, both browning and yellowing being observed. Browning can be related to the incidence of the beech leaf miner *Rhynchaenus fagi* and the fungal pathogen *Apiognomonia errabunda,* whereas yellowing appears to be primarily caused by abiotic factors (such as drought and lime-induced chlorosis). The amount of yellowing appears to fluctuate markedly from year to year and was not particularly common following the very dry summer of 1989. Preliminary examination of the data has not indicated a relationship between the amount of yellowing and the severity of the 1989 soil moisture deficit.

Relationship between crown condition and pollution

One of the reasons for undertaking the surveys is to establish whether a link exists between air pollution and tree health. It is well known that trees are affected by pollution and this has affected species selection in urban areas and around point sources of pollution (Innes, 1990b).

Table 16.5 Percentages of trees in each foliar discoloration class in a survey in Britain.

%:	Current year's needles					Older needles				
	0 0–10	1 10–25	2 25–60	3 >60	4 Dead	0 0–10	1 10–25	2 25–60	3 >60	4 Dead
(a) Needle-browning										
Sitka spruce										
1989	98	2	0	0	0	97	2	1	0	0
1988	99	1	0	0	0	89	8	3	0	0
1987	94	6	0	0	0	91	8	1	0	0
Norway spruce										
1989	99	1	0	0	0	95	4	1	0	0
1988	99	1	0	0	0	88	8	4	0	0
1987	95	4	1	0	0	94	5	1	0	0
Scots pine										
1989	93	7	0	0	0	92	7	1	0	0
1988	96	4	0	0	0	93	6	1	0	0
1987	94	5	1	0	0	92	7	1	0	0
(b) Needle-yellowing										
Sitka spruce										
1989	98	1	1	0	0	92	6	2	0	0
1988	96	4	0	0	0	93	5	2	0	0
1987	97	2	1	0	0	89	9	2	0	0
Norway spruce										
1989	98	2	0	0	0	97	2	1	0	0
1988	97	3	0	0	0	95	4	1	0	0
1987	98	2	0	0	0	98	2	0	0	0
Scots pine										
1989	97	3	0	0	0	89	8	3	0	0
1988	97	3	0	0	0	94	5	1	0	0
1987	98	2	0	0	0	96	4	0	0	0

(c) Overall needle discoloration					
%:	0–10	10–25	25–60	>60	Dead
Sitka spruce					
1989	90	7	2	1	0
1988	86	10	4	0	0
1987	88	10	2	0	0
Norway spruce					
1989	93	5	2	0	0
1988	86	9	5	0	0
1987	95	4	1	0	0
Scots pine					
1989	84	13	2	0	1
1988	90	8	2	0	0
1987	91	7	1	0	0

Table 16.5 continued.

(d) Browning and yellowing of leaves in broadleaved trees

	Browning					Yellowing				
%:	0–10	10–25	25–60	>60	Dead	0–10	10–25	25–60	>60	Dead
Oak										
1989	96	4	0	0	0	92	5	2	1	0
1988	95	5	0	0	0	92	7	1	0	0
1987	98	2	0	0	0	97	3	0	0	0
Beech										
1989	75	20	4	1	0	70	22	7	1	0
1988	72	24	4	0	0	47	35	16	2	0
1987	74	22	4	0	0	86	13	1	0	0

(e) Overall leaf discoloration in broadleaved trees in 1989

%:	0–10	10–25	25–60	>60	Dead
Oak	91	6	2	1	0
Beech	64	23	11	2	0

However, the extent to which low levels of pollution affect trees is much less clear. Trees are subject to a variety of stresses, of which air pollution is only one, and unravelling the different effects of each is complicated by interactions between stresses; for example, trees may be less or more susceptible to specific pollutants if they are already water-stressed.

Attempts have been made to relate tree condition and/or growth to various environmental parameters, including pollution (e.g. Holdaway, 1988; Innes and Boswell, 1988, 1989; Stock, 1988; Brooks, 1989; Neumann, 1989). In general, these have reached the same conclusion: that a number of factors affect tree condition or growth, and that pollution cannot be identified as the most significant or, in some cases, as even being involved. In Britain, tree health, as measured by crown condition, appears to be better in areas that experience higher levels of most forms of pollution, namely east and central England. Interpretation of this pattern is complicated by high correlations between air pollution and other environmental variables that might affect tree health. There are problems in these analyses and with the data used to construct the models; further analyses may reveal some subtle adverse relationships between air pollution and tree health that have so far been missed.

Conclusions

Despite several years of assessment, it is still not possible to state categorically whether there is a forest decline problem in Britain. From the outset, it was recognised that reliable data for at least 10 years would be required before any such conclusion could be reached. However, the data gathered so far do not suggest that forest decline is a major problem. This corresponds with data gathered in Europe, where it is now apparent that forest declines are of limited extent and probably related to local combinations of adverse environmental factors. Problems clearly exist at some sites in Great Britain, but these are localised in extent and, in most cases, the causes have been identified. There is undoubtedly pollution in Britain and increasing evidence that it has had an adverse effect on the environment. The lack of any observed effects on trees does not provide grounds for complacency, as some effects may be cumulative. Furthermore, other environmental problems such as global warming exist. It is, therefore, important that the condition of trees in the country continues to be monitored.

REFERENCES

BELANGER, R.P. and ANDERSON, R.L. (1988). *A guide for visually assessing crown densities of loblolly and shortleaf pines.* United States Department of Agriculture, Forest Service, Southeastern Forest Experiment Station Research Note SE–352.

BOSSHARD, W. (ed.) (1986). *Kronenbilder.* Eidgenössische Anstalt für das forstliche Versuchswesen, Birmensdorf.

BROOKS, R.T. (1989). An analysis of regional forest growth and atmospheric deposition patterns, Pennsylvania (USA). In *Air pollution and forest decline*, eds J. B. Bucher and I. Bucher-Wallin, 283–288, Proceedings of the 14th international meeting for specialists in air pollution effects on forest ecosystems, IUFRO project group P2.05, Interlaken, Switzerland, 2–8 October 1988. Eidgenössische Anstalt für das forstliche Versuchswesen, Birmensdorf.

COMMISSION OF THE EUROPEAN COMMUNITIES (1989). *European Community forest health report 1987–1988.* Office for Official Publications of the European Communities, Luxembourg.

HOLDAWAY, M.R. (1988). The effects of climate, acid deposition and their interaction on Lake States forests. In *Healthy forests, healthy world*, 67–71. Proceedings of the Society of American Foresters Conference, October 16–19, Rochester, New York. Society of American Foresters, Bethesda.

INNES, J.L. (1990a). *Assessment of tree condition.* Forestry Commission Field Book 12. HMSO, London.

INNES, J.L. (1990b). Plants and air pollution. In *Landscape design with plants*, ed. B. Clouston, 199–211, Heinemann Newnes, Oxford.

INNES, J.L. and BOSWELL, R.C. (1987). *Forest health surveys 1987. Part 1: results.* Forestry Commission Bulletin 74. HMSO, London.

INNES, J.L. and BOSWELL, R.C. (1988). *Forest health surveys 1987. Part 2: analysis and interpretation.* Forestry Commission Bulletin 79. HMSO, London.

INNES, J.L. and BOSWELL, R.C. (1989). *Monitoring of forest condition in the United Kingdom 1988.* Forestry Commission Bulletin 88. HMSO, London.

INNES, J.L. and BOSWELL, R.C. (1990). *Monitoring of forest condition in Great Britain 1989.* Forestry Commision Bulletin 94. HMSO, London.

INNES, J.L. and COOK, E.D. (1989). Tree-ring analysis as an aid to evaluating the effects of pollution on tree growth. *Canadian Journal of Forest Research* **19,** 1174–1189.

MAHRER, F. (1989). Problems in the determination and interpretation of needle and leaf loss. In *Air pollution and forest decline*, eds J. B. Bucher and I. Bucher-Wallin, 229–231. Proceedings of the 14th International meeting for specialists in air pollution effects on forest ecosystems, IUFRO project group P2.05, Interlaken, Switzerland, 2–8 October, 1988. Eidgenössische Anstalt für das forstliche Versuchswesen, Birmensdorf.

NEUMANN, M. (1989). Einfluss von Standortsfaktoren auf den Kronenzustand. In *Air pollution and forest decline*, eds J. B. Bucher and I. Bucher-Wallin, 209–214. Proceedings of the 14th International meeting for specialists in air pollution effects on forest ecosystems, IUFRO project group P2.05, Interlaken, Switzerland, 2–8 October 1988. Eidgenössische Anstalt für das forstliche Versuchswesen, Birmensdorf.

STOCK, R. (1988). Aspekte der regionalen Verbreitung 'Neuartige Waldschäden' an Fichte im Harz. *Der Forst und Holz* **43,** 283–286.

WESTMAN, L. and LESINSKI, J. (1986). Thinning out of the tree crown – what is hidden in that integrated measure of forest damage? In *Inventorying and monitoring endangered forests*, ed. P. Schmid-Haas, 223–228. IUFRO Conference proceedings, August 19–24, 1985, Zurich, Switzerland. Eidgenössische Anstalt für das forstliche Versuchswesen, Birmensdorf.

Paper 17
Ash dieback in Great Britain: results of some recent research

S.K. Hull, *University of Aberdeen, Department of Forestry, Cruickshank Building, St Machar Drive, Aberdeen, AB9 2UD, U.K.*

Summary

Results of a survey of dieback in hedgerow ash trees (*Fraxinus excelsior*) are reported, including the effects of different types of land use on their condition, rooting habit in arable and grassland soils and late flushing in spring 1987.

Introduction and objectives

The project investigated the possible causes of dieback in hedgerow ash trees (*Fraxinus excelsior*). This disorder has been known for many years but has only recently been examined in detail. There were two main areas of investigation: first, a survey of the hedgerow ash population in selected areas of Great Britain (reported more fully in Hull and Gibbs, 1991) and, secondly, a detailed examination of a number of hedgerow trees in relation to their surrounding environment. The unusually late flushing of ash in the spring of 1987 provided a further opportunity for research.

The background to the work was provided by a survey of ash dieback in east-central England conducted by Pawsey (1983). He found the condition to be widespread with much local variation in incidence. He suspected a link between the incidence of dieback and the intensity of agriculture in the locality, although data were not recorded on this aspect. However, two small subsequent surveys showed the incidence of ash dieback to be eight times higher in rural areas than in villages and other built-up areas.

In his report, Pawsey discussed a number of factors which might be involved in the development of ash dieback, including agricultural practices, atmospheric pollution, poor site and climatic conditions, pathogenic infections and insect infestations. He stated that initial dieback development was probably due to a complex of factors and that the condition might later be influenced by other secondary agents. The processes might take many years.

The present project was designed to build on the foundation of Pawsey's work, to collect quantitative data on ash dieback and to examine the possible influences of various factors on the development of the condition.

Summer 1987 – the main ash dieback survey

This study covered a more extensive range than that of Pawsey's ash survey, in geographical terms and in the type of information collected. The methods are described in detail in Hull and Gibbs (1991).

Two hundred 10-km squares were selected, covering all areas of Britain except the north-west of Scotland and certain other smaller areas known to have a low ash population. Field data were collected in two stages: first, in the same way as Pawsey (1983), using a car to traverse a 10-km sample unit, following secondary roads as far as possible. During the traverse, healthy and affected ash were counted separately, viewing trees up to approximately 200 m from the roadside. Affected trees were defined as having 10% or more of the crown showing dieback or with live shoots present only at the branch tips (Cooper and Edwards, 1981; Pawsey, 1983). Areas along each traverse, in which ash was common and that might form the basis of a plot from which detailed information could be efficiently collected, were noted.

Plate 17.1 *Although some of the fine twig tracery is missing, the dieback in the ash tree on the right appears to be of fairly recent origin. The tree would have been given a crown loss score of 40–49%. Northamptonshire, June 1990. (39444)*

The second stage of more-detailed data collection involved returning to the nearest area in which ash was noted to be common. Here data were gathered on foot using roads and public footpaths for access, for between 15 and 25 trees. Crown loss due to dieback (Plate 17.1) was recorded in 10% classes, i.e. trees with crown losses between 0% and 9% inclusive were entered into class 0, trees with crown losses between 10% and 19% inclusive were entered into class 1 and so on up to class 9. (For an illustration of dieback and crown loss assessment, see Figure 17.1.) The current land uses immediately surrounding each tree were described (i.e. arable, grassland, roadside, etc.) and the presence of a ditch within 5 metres of the base of the tree was noted.

Regional analysis

The data obtained from the 200 traverses, using the assessment procedures of Pawsey (1983), showed the highest levels of damage to be in the

Figure 17.1 *Distribution of ash dieback obtained from the sample plot data interpolated on to a 20 × 20 km grid; (a) percentage of affected trees, (b) percentage of severely affected trees, and (c) mean crown loss index.*

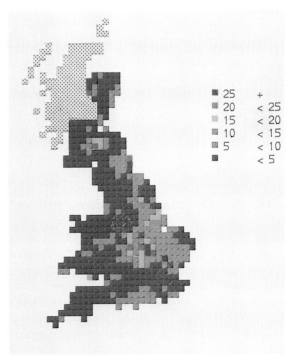

Table 17.1 Incidence of dieback in ash in relation to estimated stem diameter at breast height and position in relation to other trees.

	Diameter class				Position		
	<10 cm	10–50 cm	>50 cm	Significance	Isolated	Grouped	Significance
Number of trees sampled	215	3832	407		1515	2939	
Number of trees affected	2	665	166	***	393	440	***
%	1	17	41		26	15	
Number of trees severely affected	0	229	85	***	176	138	***
%	0	6	21		12	5	
Crown loss index (see text below)	0	5	13	***	8	4	***

*** significant at 0.1% level ($P<0.001$).

Table 17.2 Incidence of dieback in ash in relation to surrounding land use.

	Urban	Arable only	Grassland only	Some arable	Some grassland	Some roadside	No roadside
Number of trees sampled	448	297	786	1339	2338	2050	2026
Number of trees affected	49	112	77	415	331	465	252
%	11	38	10	31	14	23	12
Number of trees severely affected	15	29	27	169	124	203	72
%	3	10	3	13	5	10	4
Crown loss index	3	8	3	9	4	7	3

east of the country, in particular around the south-east Midlands. Within this area, the distribution of dieback was similar to that found by Pawsey (1983), although maximum values were rather less.

Data from the more detailed survey plots were summarised in three ways, in terms of (i) the proportion of 'affected' trees, i.e. the proportion with at least 10% crown dieback; (ii) the proportion of 'severely affected' trees, i.e. the proportion with at least 30% crown dieback; and (iii) a crown loss index. The latter was obtained using data from the dieback classes, whereby a tree in class 0 was taken to have a crown loss of 0, one in class 1 a crown loss of 10, and so on. The scores from all the trees under consideration were then averaged to produce a mean crown loss index. The indices presented in the tables are rounded to the nearest whole number.

Overall, the incidence of dieback in the sample of 4454 trees was 19%, while the proportion of severely affected trees was 7%. The distribution of affected ash trees interpolated from sample data (Figure 17.1a) was similar to that indicated by Pawsey's assessment method: a pattern of widespread dieback with a concentration of damage in the south-east Midlands and with some localised dieback in other, generally easterly, locations. The distribution of severely damaged ash (Figure 17.1b) and the mean crown loss (Figure 17.1c) were also similar.

Tree size and position in relation to other trees

Table 17.1 shows the health of ash in relation to estimated stem diameter at breast height; for trees under 10 cm diameter, dieback is negligible. With increasing diameter, there was an increase in the incidence of damaged and severely damaged trees. The incidence of damage and severe damage in isolated trees was greater than that in trees in groups.

Local site factors

Sixteen types of land use were encountered during the survey (Table 17.2 shows the most important ones) and when considering individual trees there could be many combinations.

Table 17.3 Incidence of dieback in ash in relation to land use and presence (P) or absence (A) of ditches.

	Arable only			Grassland only			Some arable			Some grassland			Some roadside			No roadside		
	Ditch			Ditch			Ditch			Ditch			Ditch			Ditch		
	P	A	Sig	P	A	Sig	P	A	Sig	P	A	Sig	P	A	Sig	P	A	Sig
Number of trees sampled	45	144		13	745		228	852		143	2057		307	1743		78	1948	
Number of trees affected	16	54	ns	2	67	ns	96	227	***	41	259	***	108	357	***	29	223	***
%	36	38		15	9		42	27		29	13		35	20		37	11	
Number of trees severely affected	7	14	ns	0	24	ns	50	89	***	17	95	***	55	148	***	12	60	***
%	16	10		0	3		22	10		12	5		18	8		15	3	
Crown loss index	10	8	ns	2	2	ns	15	7	***	8	3	***	12	6	***	12	3	***

n.s. not significant
*** $P < 0.001$

Trees in urban areas were among the healthiest found during the survey, with 11% affected and only 3% severely affected by dieback. This contrasts with data from trees in rural locations: 20% of trees affected and 7% severely affected. Trees where only arable land was recorded around them had a high incidence of damage as did trees where arable and any other land use was recorded. Trees next to grassland showed little dieback. The categories 'some arable' and 'some grassland' were analysed because of the small numbers of trees in the more restricted categories. They include, respectively, 'arable only' and 'grassland only' trees but they also include data on the much larger numbers of trees where arable or grassland was present with some other land use.

Trees were examined in relation to the presence or absence of a nearby road: the incidence of affected and severely affected trees was approximately twice as high in roadside trees as in trees remote from the road.

These land-use effects were examined further by subdivision into trees with or without a ditch within 5 metres of the base of the tree (Table 17.3). In all the cases examined, except 'arable only' where there appeared to be no relationship, and 'grassland only', where an effect was apparent but was not statistically significant, the presence of a ditch was associated with a significant increase in the proportion of damaged trees. In many instances, the proportion of severely affected trees within 5 metres of a ditch was more than twice that where there was no ditch.

Selected environmental variables

As the project was initiated partly as a result of concern for the effects of pollution on trees, the data were analysed in relation to some pollution variables as well as to more general environmental factors. Pollution data prepared by Warren Spring Laboratory and used in the 1987 Forestry Commission Forest Health Survey (Innes and Boswell, 1988) were used with data for annual rainfall, summer rainfall and soil moisture deficit from published maps (Birse and Dry, 1970; Meteorological Office, 1979a, b; Soil Survey of England and Wales, 1980).

The nationwide pollution data, provided on a regular grid basis, were interpolated by computer to provide a value for each of the 200 plots visited in the survey. These data with those for the other environmental variables were then used with the mean crown loss index in each plot in a simple correlation analysis. For an analysis of the fewer trees associated with particular land uses, e.g. 'arable only' trees, data were pooled into larger units of 50 × 50 km squares.

Table 17.4 shows any statistically significant correlations that were found for pollutant, rainfall and soil moisture variables used in this study.

Table 17.4 Correlation analysis between plot mean crown loss and environmental variables [a] in ash and oak data sets.

	Ash			Oak	
	All ash	Arable only	Some arable	All oak	Grassland only
S deposition	—	—	—	—	—
H concentration	0.1935 **	0.5776 **	0.2737 *	—	—
H deposition	—	0.5130 **	—	—	—
NH_4 concentration	0.1603 *	0.3730 *	—	—	—
NH_4 deposition	—	—	—	—	—
Non-marine SO_4 concentration	0.1504 *	0.5487 **	—	—	—
Non-marine SO_4 deposition	−0.1451 *	0.5001 **	—	—	—
NO_3 concentration	0.1941 **	0.4138 *	—	—	—
NO_3 deposition	—	—	—	—	—
Total SO_4 concentration	—	0.5451 **	—	—	—
Total SO_4 deposition	−0.2496 **	0.4111 *	−0.2817 *	—	−0.3615 *
SO_2 concentration	—	0.4093 *	0.3461 *	—	—
Mean annual rainfall	−0.3451 ***	—	−0.3671 **	−0.2362 *	—
Mean summer rainfall	−0.3197 ***	—	−0.3306 *	−0.2360 *	—
Mean soil moisture deficit	0.2410 **	—	—	—	—
Number of trees sampled	4454	297	1339	1022	233
Number of plots	200	30	53	121	32

[a] For details of sources of environmental data, see Hull and Gibbs (1991).
Note: Coefficients listed only where statistically significant: * $P < 0.05$, ** $P < 0.01$, *** $P < 0.001$.

When the data for 'all ash' trees were examined, a number of significant relationships emerged. Significant negative correlations with damage were found with mean annual rainfall and with mean summer rainfall ($r = -0.35$ and $r = -0.32$ respectively), crown loss increasing with decreasing rainfall. Significant correlations were also found with several pollutants. The strongest, also being negative, was with total wet deposited sulphate ($r = -0.25$). A significant positive correlation was obtained with soil moisture deficit ($r = 0.24$).

For trees with 'arable only' land use, there was no correlation with rainfall, but damage correlated well with hydrogen ion concentration and deposition and total sulphate concentration ($r > 0.5$ for all the above).

For 'some arable' trees the situation was more like that found with the population as a whole than with the 'arable only' population. Significant negative correlations were found with rainfall ($r = -0.37$ for annual rainfall and $r = -0.33$ for summer rainfall) and total sulphate deposition ($r = -0.28$). Sulphur dioxide concentration was significantly positively correlated with dieback ($r = 0.35$), as was hydrogen ion concentration ($r = 0.27$). Total sulphate deposition was significantly negatively correlated ($r = -0.28$) with damage. No significant relationships were found with the 'grassland only' data.

Factors affecting ash dieback

Hull and Gibbs (1991) discuss various features of current agricultural practice, in particular of arable farming, that could have an adverse effect on adjacent ash trees. Stubble burning was considered unlikely to be of major importance but ploughing and other tillage operations together with soil compaction could have great potential for causing damage to trees through effects on the rooting system. Soil compaction, as a result of animal and vehicle movements, may also be a factor on grassland. The use of herbicides and fertilisers on crops could damage adjacent long-lived trees in unexpected ways, e.g. by increasing their susceptibility to adverse biotic and abiotic agents of disease (Sinclair *et al.*, 1987; Van Breemen and Van Dijk, 1988).

The poor health of trees next to roads compared with those remote from roadsides is not easy to interpret, bearing in mind that most of the roads were minor ones. The amount of fumes produced by traffic on such roads would be relatively small and there would have been little use of de-icing salt. Service installation is an uncommon event along this type of road, although roadside ditch maintenance by local authorities may be of some importance. Away from roadsides, ditch management may be a factor to consider.

The association with particular pollution variables did not point conclusively to their importance, although the strongest positive correlations were with the 'arable only' trees, i.e. those worst affected by dieback. In such situations pollution may exacerbate the effects of other kinds of stress. Unfortunately, at the time of the study, suitable data on the distribution of peak ozone concentrations were not available for use in a correlation analysis; there is some indication that *Fraxinus* may be intolerant to this pollutant (Karnosky and Steiner, 1981).

Root distribution

Following the completion of the survey it was clear that the differences in dieback on arable and grassland needed further investigation. Work was carried out during summer 1988 at sites with a long history of arable or grassland farming where ash trees were present. Soil pits were dug at the crown edge of open-grown trees so that one side of the pit was carefully excavated to expose the roots occurring in the vertical plane. Numbers of roots and their sizes were recorded on a systematic grid of 5 × 5 cm squares over a total width of one metre and to a depth of 75 cm.

The most striking feature to emerge from the data (Table 17.5) is that trees on arable sites had far fewer roots than trees on grassland sites, the difference being particularly marked when only the upper 20 cm of soil was considered. On grassland sites, 31% of all roots occurred in this top layer of soil compared with just over 5% in arable soils.

These results can be readily understood in terms of soil tillage processes, since ploughing usually inverts the soil to depths of 15–20 cm. It may be significant for ash dieback that there appears to be no compensation at lower depths for these 'missing' roots.

At five sites no roots were recorded; at four of these sites drainage ditches (50–90 cm deep) were present along the field edge between the tree and the pit profile, suggesting that ditches effectively prevent tree roots from exploiting substantial areas of soil that would otherwise be available to them. This may have an important influence on the susceptibility of such a tree to ash dieback.

Exceptionally late flushing of ash in 1987 – an unscheduled research opportunity

Typically, ash is one of the last trees to come into leaf in the spring, rarely flushing before May (Edlin, 1985). However, in spring 1987, flushing was exceptionally late and this gave rise to concern over the possible initiation of a new phase of ash dieback. As research into the problem was by this time underway, the phenomenon came at an opportune moment and a study of trees in 12 locations was initiated.

A full description of the work is given in Hull (1991). Nineteen trees were visited in June 1987, when special attention was given to recording any damage to buds caused by fungi, particularly *Nectria* sp., or by the mining larvae of the ash bud moth (*Prays fraxinella*), since

Table 17.5 Distribution with depth of ash roots in arable and grassland soils.

Depth in soil profile (cm)	Mean number of roots per horizontal metre	
	Arable	Grassland
0–10	1	344
10–20	64	951
20–30	297	752
30–40	353	520
40–50	240	381
50–60	203	405
>60	190	710

Plate 17.2 *Ash tree showing recovery from dieback. The foliage is carried almost entirely on secondary growth. The current crown loss score of 20–29% is probably lower than it would have been in the past. Northamptonshire, June 1990. (39429)*

these organisms have been postulated as being important in contributing to ash dieback (Pawsey, 1983). Twigs, some with obvious fungal infection and others showing signs of recent dieback, were brought back to the laboratory for fungi to be isolated in artificial culture.

A visit to the same trees in high summer 1987 revealed that all had made a full recovery; their crowns were fully foliated and no abnormal amounts of new dieback were detected. The trees had flushed normally when they were reassessed in June 1988.

As might be expected, the data for 1987 and 1988 showed a large difference in the number of buds flushing to produce healthy shoots. Hull (1991) suggested that this may, in part, have been caused by a period of unusually warm spring weather (Meteorological Office, 1987) having a disruptive effect on the breaking of dormancy (Barnola et al., 1986). There was a greater proportion of dead buds in 1987 than in 1988. The main difference was accounted for by buds killed by unknown causes, there being no difference between those killed by bud moth or fungal infection. This effect may be due to an improvement in the perception of damage during the course of the investigation. Thus, very small entry holes made by bud moth larvae may have been overlooked in the first year.

Of the fungi isolated from buds and twigs, the two most common were *Fusarium lateritium*, usually associated with buds, and *Nectria coccinia*, from the dead–live junction of dying twigs. *N. galligena*, usually found in association with ash canker (Boa, 1981), was not found.

This late-flushing episode is not unique: *The Journal of Forestry and Estates Management* in 1878 chronicles a similar phenomenon (Anon., 1878). However, records at the Archive Section at the Meteorological Office revealed no similarities to the warm temperatures recorded in 1987 prior to the late-flushing episode.

Conclusions

The need for future research

This project has substantially extended our knowledge of the distribution and severity of ash dieback in Great Britain. It is clear that it is a widespread condition, particularly common in the east Midlands, the area where it first aroused attention almost 30 years ago. It is essential that the components of current agricultural practice that are damaging to trees should be identified and, to this end, there is a requirement for more-detailed studies of tree pathology. Assessments of crown condition in conjunction with growth patterns and known field history would provide useful data. Experimental work in which individual trees are monitored at sites where various agricultural treatments have been carried out should also be undertaken.

Tree declines or diebacks result from a complex interactive set of factors, biotic and abiotic, producing a gradual general deterioration in health and often ending in tree death (Houston, 1981). However, if one or more of these factors moderate in intensity, it is not uncommon for affected trees to enter a recovery phase. That this can happen with ash dieback is clearly indicated by the many trees in which the growth of secondary shoots has led to a reconstruction of a sizeable crown (Plate 17.2). Detailed studies of secondary shoot growth may be useful in identifying key factors involved in the development of dieback.

The need for future research funding is particularly important in view of the current interest in planting new woodlands on former agricultural land and in the development of 'agroforestry' systems of land management. Ash would seem to have much to offer in such schemes but uncertainty regarding the causes of dieback must cast some doubt over its future role.

ACKNOWLEDGEMENTS

Funding for the project was provided by the Department of the Environment (Air and Noise Division) and work was carried out jointly with the Forestry Commission and the University of Aberdeen. I would like to thank Dr Richard Pawsey; all my colleagues in the Pathology Branch of the Forestry Commission at the Forest Research Station, Alice Holt Lodge, especially Dr John Gibbs; David Mobbs for statistical support; and Marie-Anne Campbell and many others for field assistance.

REFERENCES

ANON. (1878). Editorial notes. *The Journal of Forestry and Estates Management* **2** (July), 148.

BARNOLA, P., LAVARENNE, S. and GENDRAUD, M. (1986). Dormance des bourgens apicaux du frêne (*Fraxinus excelsior* L.): évaluation du pool des nucléosides triposphates et éventail des températures actives sur le débourrement des bourgeons en periode de dormance. *Annales des Sciences Forestières* **43**, (3), 339–349.

BIRSE, E.L. and DRY, F.T. (1970). *Assessment of climatic conditions in Scotland 1. Based on accumulated temperature and potential water deficit*. Macaulay Institute for Soil Research, Aberdeen.

BOA, E.R. (1981). *Ash canker disease*. Ph.D. thesis (unpublished), Department of Plant Sciences, University of Leeds.

COOPER, J.I. and EDWARDS, M.L. (1981). Viruses of trees. In *Forest and woodland ecology,* eds F.T. Last and A.S. Gardner, 143–145. Institute of Terrestrial Ecology, Symposium No. 8, Cambridge.

EDLIN, H.L. (1985). *Broadleaves.* Forestry Commission Booklet 20. (Revised by A.F. Mitchell.) HMSO, London.

HOUSTON, D.R. (1981). *Stress triggered tree diseases. The diebacks and declines.* United States Department of Agriculture, Forest Service, Booklet NE–INF–41–81.

HULL, S.K. (1991). A study undertaken in southern Britain to investigate the late flushing of hedgerow ash in spring 1987. *Forestry* **64,** 189–198..

HULL, S.K. and GIBBS, J.N. (1991). *Ash dieback – a survey of non-woodland trees.* Forestry Commission Bulletin 93. HMSO, London.

INNES, J.L. and BOSWELL, R.C. (1988). *Forest health surveys 1987. Part 2: analysis and interpretation.* Forestry Commission Bulletin 79. HMSO, London.

KARNOSKY, D.F. and STEINER, K.C. (1981). Provenence and family variation in response of *Fraxinus americana* and *F. pennsylvanica* to ozone and sulfur dioxide. *Phytopathology* **71,** 804–807.

METEOROLOGICAL OFFICE (1979a). *Average annual rainfall (mm). Period 1941–1970.* HMSO, London.

METEOROLOGICAL OFFICE (1979b). *Average rainfall (mm). Period 1941–1970. Summer half year April to September.* HMSO, London.

METEOROLOGICAL OFFICE (1987). *Monthly weather reports (for April and May 1987)* **104,** Nos. 4 and 5, respectively. HMSO, London.

PAWSEY, R.G. (1983). *Ash dieback survey. Summer 1983.* Comonwealth Forestry Institute, Occasional Paper 24.

SINCLAIR, W.A., LYON, H.H. and JOHNSON, W.T. (1987). *Diseases of trees and shrubs.* Cornell University Press.

SOIL SURVEY OF ENGLAND AND WALES (1980). *Average maximum potential cumulative soil moisture deficit.* Soil Survey of England and Wales, Harpenden.

VAN BREEMEN, N. and VAN DIJK, H.F.G. (1988). Ecosystem effects of atmospheric deposition of nitrogen in The Netherlands. *Environmental Pollution* **54,** 249–274.

Discussion

S.R.M. Jones (Richard Loader Tree Care)
Is there any relation between ash dieback and ploughing?

S.K. Hull
No trials have been carried out to investigate such a relationship, but could be incorporated in future work.

A.D. Bradshaw (Liverpool University)
Ash dieback requires long-term experimentation, so why not plant ash trees on which destructive treatments can be applied? Good systematic experiments on weed control established the effects of weed competition on successful tree establishment, so the same principles could be applied to determining the cause of ash dieback.

J.N. Gibbs (Forestry Commission)
Experiments of this nature have already been suggested, but have not been financed.

A. Coker (Department of Transport)
Roadside ditches tend to concentrate nutrient salts. Has any difference between ash growing along roadside ditches and elsewhere been noted?

S.K. Hull
Differences in dieback between roadside and other sites are not yet fully understood and require further investigation.

Disorders of amenity trees

Paper 18
De-icing salt damage to trees and shrubs and its amelioration

M. C. Dobson, *Forestry Commission Research Station, Alice Holt Lodge, Farnham, Surrey, GU10 4LH, U.K.*

Summary

De-icing salt in run-off and spray can significantly alter the health of roadside trees and shrubs. Symptoms of damage range from depression of growth, leaf scorch and defoliation to limb dieback and ultimately death. To minimise damage to roadside vegetation, tolerant species should be planted in areas of high risk. Further measures for reducing damage from salt include reducing salt applications, using alternative de-icers (e.g. calcium-magnesium acetate), leaching of contaminated soil, and the use of soil ameliorants such as gypsum (calcium sulphate) and low-nitrogen fertilisers.

Introduction

Rock salt, sodium chloride, has been used in increasing quantities since the Second World War throughout Europe and the USA for minimising the danger to motorists and pedestrians from icy thoroughfares. In Britain, rates of application have increased sharply so that, for example, applications during the winter of 1979/80 were more than twice those in the appreciably more severe winter of 1962/63 (Figure 18.1). During hard winters total application of salt may exceed 3 kg m^{-2} of road in Britain (Davison, 1971) and 5 kg m^{-2} in some areas of Europe (Defraiteur and Schumaker, 1988). Following precipitation, this salt leaves the road either as run-off or as spray whipped up by fast-moving traffic. Salt spray may be intercepted by the aerial parts of roadside vegetation, and salt in run-off tends to accumulate in soil close to the roadside. The resulting damage to vegetation ranges from marginal leaf necrosis in broadleaves and tipburn in conifers, to premature defoliation, limb dieback and even death. Jordan (in 1971), cited by Flückiger and Braun (1981), estimated that de-icing salt applications were directly responsible for the death of 700 000 trees a year in Western Europe.

In Britain, periodic severe crown dieback of London plane (*Platanus × hispanica*) has caused concern for a number of years (Plate 18.1). Gibbs and Burdekin (1983) showed that much of the damage could be attributed to de-icing salt applications during severe winters (see Figure 18.1). However, the scale and distribution of damage to London plane has not been assessed and further work on commonly planted roadside trees has not been carried out. Therefore, although it is recognised that de-icing salt can injure roadside vegetation, the severity and distribution of damage to amenity species in this country is almost entirely unknown.

Symptoms of salt damage

Before the extent of damage to vegetation from de-icing salt can be assessed, one must be able to recognise and diagnose salt damage. A description of symptoms has been summarised from the literature, based mainly on observations in the USA, Canada and mainland Europe because of the lack of information from Britain.

Symptoms differ somewhat for broadleaved-deciduous and evergreen species and also depend on whether damage is caused by salt spray or salt contamination of soil.

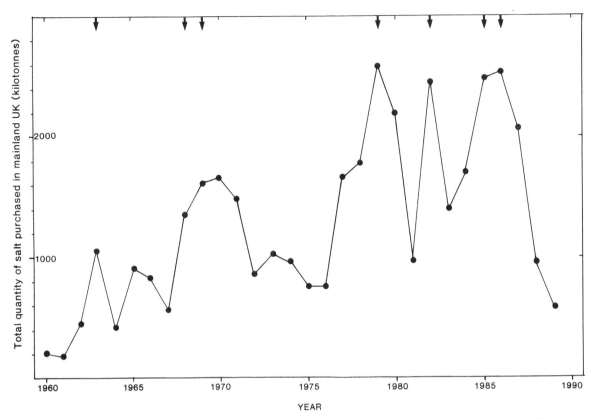

Figure 18.1 *Estimates of total quantity of de-icing salt purchased annually in mainland Britain in 1960–1989. In the early 1960s, Highway departments changed from using salt/abrasive mixtures to pure rock salt and this may account for some of the increase in salt usage. Arrows represent years when significant crown dieback of London plane was reported (Gibbs and Burdekin, 1983; Forestry Commission records). Data were supplied by ICI.*

Deciduous species

Soil salt. In contrast to evergreen species, in which salt damage becomes apparent during winter, symptoms on deciduous species only become evident once buds begin to open in the spring. The symptoms associated with soil-salt toxicity have been identified mainly for broadleaved trees along city streets (see also Sucoff, 1975):

i. General reduction in growth, often proportional to the chloride content of the shoots. This symptom is not specific to salt damage but always occurs if other symptoms are present.

ii. Marginal necrosis (browning) of leaves (Plate 18.2). Often only present on individual limbs, frequently those facing the road. Necrosis begins as a yellowing of the leaf margin which subsequently turns brown. There is usually a sharp boundary between necrotic and healthy tissue. Necrosis spreads in the interveinal tissue and may sometimes be accompanied by wrinkling and curling. This is the symptom most characteristic of salt damage and the one most easily used for diagnosis.

iii. Premature autumnal colouring, followed by early leaf fall. This may occur before or after symptom ii.

iv. Small, yellowish leaves, few in number, giving the crown a generally unhealthy and thin appearance.

v. Post-flushing dieback (Plate 18.3). A symptom that has, until recently, only been noted for London plane (Gibbs and Burdekin, 1983). Partially expanded

leaves wither and die shortly after flushing. These small dead leaves may remain attached to the stems throughout the season. (Symptoms of salt damage in plane may be confused with the disease anthracnose *(Gnomonia platani);* Gibbs, 1983.)
vi. Stem lesions (Plates 18.4 and 18.5).
vii. Twig dieback followed by limb dieback (Plate 18.5). This leads to a reduced branch structure and a thinner crown.
viii. Tree death.

Damage often develops progressively over a number of years but in severe winters with heavy salt applications all the above symptoms may appear in a single year.

Salt spray. Damage from salt spray may be found beside any salted road with fast-moving traffic but is often most noticeable along motorways and trunk roads. It is uncommon for spray damage to be seen in towns and cities because of slow traffic movement. The symptoms are as follows:
i. Dieback of the previous year's twig growth beginning at the twig tip.
ii. Death of buds. This occurs with salt entering at leaf scars. The internodes may also show dead spots and examination of the cambial tissue may reveal almost total browning.
iii. Release of lateral buds. Because of the death of apical buds, lateral buds on wood more than one year old are released from dormancy often giving a 'witches broom' appearance.
iv. Delay of leafing out in spring of up to 3 weeks.

In contrast to damage from salt in the soil, marginal necrosis is unlikely to occur as a result of salt spray because chloride (Cl-) concentrations in the developing leaves rarely, if ever, reach toxic levels.

Evergreen species

Soil salt. Symptoms associated with damage from soil salt for conifers are less well documented than for salt spray symptoms but include:

i. death of buds or delay in flushing of up to 3 weeks; and
ii. yellowing or browning of needles during the season they emerge from the bud.

Salt spray. A large amount of data on the effects of salt spray has been gathered for pines. Symptoms for fir and spruce are similar. For pine, symptoms are as follows:
i. Tipburn on needles one or more years old (Plate 18.6). (Data from North America indicate that damage only becomes evident after the spring thaw when temperatures begin to rise above freezing. However, this may not be the situation in Britain, where temperatures fluctuate considerably during winter.) The tips of the needles turn yellow, then bronze and eventually become necrotic. There is usually a clear demarcation between the necrotic and healthy tissue. At an early stage in this process, yellow bands sometimes appear across the needles and these may exude resin (Spotts *et al.,* 1972).
ii. Progressive necrosis of 2–3-year-old needles, browned in previous years. Necrosis continues towards the base and needles that are more than half browned tend to fall off.
iii. Chlorotic fleck of needles. The green part of the needles may become flecked with tiny bleached spots which often disappear later in the season. These flecks help to distinguish salt damage from other types of needle browning (Sucoff, 1975).

Buds are rarely affected by salt spray, and flush normally in the spring; new needles remaining green and healthy throughout their first year. This tends to mask the older needles, giving the appearance of recovery, but where needle shedding is considerable trees have noticeably thinner crowns.

Damage from salt in the soil should also be suspected when trees sheltered from salt spray show as much damage as those exposed to it.

The situation in other conifers is broadly similar. Foster and Maun (1978) have shown that

Plate 18.1 Almost complete defoliation of a mature London plane caused by salt. The tree was at the edge of an artificially surfaced football pitch which was salted once in March 1989 at 30–40 g of salt m^{-2}. Two nearby trees were also badly affected. (D. Thorogood)

Plate 18.2 Severe marginal necrosis (left) in a leaf taken from a 90% defoliated London plane tree, in London, compared with a leaf from a healthy tree nearby (right). The chloride content was 4.4–5.1% of the dry weight in leaves from the damaged tree and 0.37–0.43% in healthy leaves. (M. Levy)

Plate 18.3 Post-flushing dieback caused by de-icing salt in a London plane on the Embankment in London. Photograph taken in May following the severe winter of 1986/87 (see also Figure 18.1). (J. N. Gibbs)

Plate 18.4 Stem lesion on a salt-damaged London plane extending several metres up the trunk. (D. Thorogood)

Plate 18.5 Salt-damaged London plane in Kensington. Extensive dieback and pruning of dead wood has left this tree with a severely reduced crown. Note the extensive stem lesions which have almost girdled the tree. Because of its unsightliness and ill health, this tree will eventually have to be removed. In London alone, hundreds of mature planes are killed outright following winters with heavy salt applications. Quite apart from the loss of amenity value, the removal costs run into hundreds of thousands of pounds. (M. C. Dobson)

Plate 18.6 Tipburn of Pinus radiata beside the A31 trunk road in Hampshire. The chloride content of the damaged needles was 1.6% of the dry weight compared with 0.3% in undamaged needles. This damage occurred during the 1989/90 winter which was extremely mild. Trunk roads in Hampshire were only salted approximately 20 times during that winter. (39297)

salt spray causes yellowing and necrosis of shoot tips in Eastern white cedar *(Thuja occidentalis)*. Necrosis extends basipetally and the demarcation between the necrotic and healthy tissue is sharp. Symptoms are similar for soil-applied salt, but yellowing also occurs at the bases of secondary and tertiary branches and progresses acropetally and basipetally. Bernstein *et al.* (1972) also showed that soil salting could cause necrotic shoot tips and basal shoot necrosis in *Thuja orientalis* and *Juniperus chinensis*.

There are few reports of symptoms for broadleaved evergreens. Dirr (1975) showed that salt caused marginal necrosis of English ivy *(Hedera helix)*. Some species may show bronzing rather than necrosis of leaf margins or tips (e.g. Chinese privet, *Ligustrum lucidum*) and others may show bronzing in addition to scorch (e.g. Chinese shrubby holly, *Ilex cornuta*) (Bernstein *et al.*, 1972). Other species may show milder symptoms, such as slight marginal necrosis, and shed their leaves before stronger symptoms develop. An example of this is pittosporum *(Pittosporum tobira)*, which, in a controlled experiment, showed severe leaf drop in response to soil-applied salt, leaving only rosettes of leaves at the end of bare stems (Bernstein *et al.*, 1972).

General patterns of injury

In addition to the symptoms described above, location of damaged vegetation and distribution of injury within a plant are also valuable guides when assessing the possible contribution of de-icing salt to damage of roadside trees and shrubs. The following are general injury patterns that have been identified when soil salt is the primary cause of injury, and may apply equally to salted paths and pavements as to roads.

1. Trees within 5 m of the roadside are worst affected, but there is frequently a distinct injury gradient with distance from the road. (However, where roots penetrate drains carrying salty run-off, damage may occur at a considerable distance from the roadside.)
2. Trees on the downhill side of the road suffer more damage than those on the uphill side.
3. Trees planted in depressions or with a depression around their base (e.g. where planting soil has settled) suffer more damage than trees in raised planting sites.

Where salt spray is the main cause of damage the following patterns of injury can be seen.

4. Trees on the downwind side of a highway show greatest injury.
5. Injury is greatest on the side of the tree facing the road. (Trees are often misshapen and one-sided in appearance because of the death of buds and branches facing the road.)
6. Trees sheltered from spray, e.g. by a fence or wall between trees and road, lack injury symptoms.
7. Upper branches, above the spray-drift zone, are not injured or are less injured than lower ones.
8. Flowers may only open on the side of the tree facing away from the road, because flowering buds are more sensitive to salt spray than leaf buds.
9. Where deep snow lies for significant periods, parts of the tree below the snow line are not injured.

Mechanisms of action

Salt causes damage by two mechanisms: directly, by altering plant metabolism and, indirectly, by altering the structure and nutritional status of soils. Sodium (Na^+) and chloride (Cl^-) may be absorbed through leaves (salt spray) or through roots (soil salt); Cl^- appears to be the more toxic ion and is particularly mobile once inside the plant. It moves upwards with the transpiration stream and accumulates in the tips of twigs and leaves. When concentrations become critical, plasmolysis of cells occurs and necrosis of leaves and shoots may result (Plates 18.2 and 18.6). If leaf symptoms are present, concentrations of Na^+ and Cl^- in the leaf are usually greater than normal. Analysis for these ions is therefore a useful diagnostic technique when salt damage is suspected.

Foliar Cl^- concentration is a more reliable

indicator of damage than foliar Na^+. Concentrations of Cl^- associated with foliar injury differ from species to species, for example, in London plane, leaf injury does not occur until foliar concentrations exceed approximately 1% of the dry weight of the leaf, but in Norway maple *(Acer platanoides)* injury may occur at concentrations as low as 0.2%. Information on Cl^- and Na^+ concentrations associated with foliar injury for a wide range of species has been summarised by Sucoff (1975) and more recently by Dobson (1991).

In the soil, Na^+ is the more damaging of the two ions; it is readily taken up onto cation exchange sites and displaces other nutritionally important ions such as Ca^{2+}, Mg^{2+} and K^+. Sodium is also taken up by plants in preference to other cations and this further exacerbates nutritional deficiencies. For example, French (1959) has shown that Ca^{2+} uptake may be reduced by up to 40% in sodium-affected soils. Besides impoverishing the soil, increased concentrations of sodium lead to increased pH and a breakdown of the soil structure, which enhances soil compaction and inhibits root penetration. Chloride similarly competes with other anions, particularly phosphate but also sulphate and nitrate, resulting in anionic nutrient imbalance. However, Cl^- is less persistent than Na^+ in the soil and is leached out more rapidly, being repelled by the negatively charged soil colloids.

Species tolerance

Although no plants are totally immune to salt injury, some species are more tolerant of salt than others. One long-term method of reducing salt damage to roadside plants is therefore to select species with high salt tolerance, for planting in areas where salt damage is likely to occur. However, there are several difficulties in producing definitive lists of salt-tolerant species. Species can only be ranked in order of relative tolerance, and ranking therefore depends on whether a species is compared with a more or less tolerant species. Relative tolerance may change according to season, soil type and climate; mature trees may be more or less susceptible than juvenile trees; and trees tolerant of salt in their rooting zone may not be tolerant to salt spray (e.g. oak) and *vice versa*. Taking into account some of these factors, the reported salt tolerance of selected amenity species has been evaluated (Table 18.1; for more detailed information see Dobson, 1991). Although this information should be regarded with caution, there are enough consistent data

Table 18.1 Relative tolerance to de-icing salt of selected tree and shrub species common in the British roadside environment, compiled from the literature. Species within a group are ranked in order of decreasing tolerance.

Tolerant	Moderately tolerant	Intermediate	Moderately susceptible	Susceptible
Populus alba[1]	Quercus robur[1]	Betula pendula[1]	Corylus colurna[4]	Sorbus aucuparia[1]
Sophora japonica[1]	Gleditsia triacanthos[1]	Platanus × hispanica[2]	Pinus sylvestris[1]	Fagus sylvatica[1]
Populus canescens[1]	Fraxinus angustifolia[3]	Ilex aquifolium[3]	Alnus incana[2]	Carpinus betulus[1]
Salix fragilis[1]	Quercus rubra[4*]	Ligustrum lucidum[3]	Aesculus hippocastanum[4]	Rosa canina[1]
Pinus nigra[1]	Rosa rugosa[4*]	Salix caprea[2]	Tilia cordata[4]	Crataegus crus-galli[2]
Robinia pseudoacacia[1]	Ailanthus altissima[3]	Sorbus aria[3]	Populus nigra 'Italica'[3]	Aesculus × carnea[3]
Salix alba[1]	Prunus avium[3]	Acer platanoides[4]	Tilia platyphyllos[2]	Cornus florida[3]
Quercus petraea[2]		Acer saccharinum[1]	Acer pseudoplatanus[1]	Tilia × euchlora[3]
Tilia tomentosa[3]		Fraxinus excelsior[4]	Cornus mas[1]	
Prunus cerasifera[3]		Prunus padus[4]		
Ginkgo biloba[3]				
Rosa pimpinellifolia[3]				

[1] Confident classification; [2] Reasonably confident classification; [3] Insufficient data for reliable classification; [4] Conflicting data.
* Tolerant of soil salt but not of salt spray.

to suggest that species such as *Populus alba, Pinus nigra, Robinia pseudoacacia,* and *Sophora japonica* are relatively salt tolerant. At the other extreme, there is also sufficient information to indicate that *Sorbus aucuparia, Fagus sylvatica, Carpinus betulus* and *Rosa canina* are highly susceptible to salt damage; planting of these species in areas where salt is likely to be a problem should be avoided.

Prevention

It would be impractical and unwise to suggest that road de-icing should be discontinued as this would not only cause traffic delays but also place the lives of road users at greater risk. However, the two currently specified rates at which salt should be applied to roads (10 g m^{-2} for precautionary salting and up to 40 g m^{-2} before or during severe wintry weather; Department of Transport, 1987) are regularly exceeded in Britain (Audit Commission, 1988). This is partly because much of the equipment used for salt spreading, especially that used by Local Authorities, is old and worn. Nevertheless, regular calibration and checking of equipment, and training of vehicle operators, could result in reductions in the use of salt of the order of 20–25%.

Another approach to reducing salt applications is through improved prediction of icy conditions (Parmenter and Thornes, 1986). Frequently, salting operations are carried out unnecessarily because predicted icy conditions do not arise. Use of temperature sensors linked in with the Meteorological Office 'Open Road' service, which has a correct frost prediction rate of 87%, can minimise unnecessary salting. Thermal mapping used in conjunction with these prediction systems could result in reductions in costs, and salt usage, of up to 30%. Considering that county-wide salting for a single night costs about £20,000, this represents a considerable saving.

Improved storage of salt, better control of salt applications and improved prediction of icy conditions could result in a total reduction in the annual consumption of de-icing salt by up to 60%, with an annual saving in the cost of salt of approximately £21 million (Audit Commission, 1988).

Alternatives

At present there is no cheap, effective, non-toxic alternative to salt. Optimisation of salt applications is therefore probably the most effective way of reducing de-icing salt damage. However, some alternatives have been tried and are currently in use for specific areas. For example, because of concern about corrosion, urea is used instead of salt on some elevated sections of the M6 and the Severn Bridge. This is non-corrosive but is a less effective de-icer and costs about ten times as much as salt. High concentrations of urea may also be detrimental to trees because they may make them more susceptible to frost injury. In airports, glycols are used on runways because of worries about corrosion of aeroplanes. However, both urea and glycol may increase biological oxygen demand of rivers receiving run-off and this is of concern to River Authorities.

A more promising alternative is calcium magnesium acetate (marketed by BP Chemicals as 'Clearway CMA De-icer'), which has a similar de-icing capability to salt. In preliminary tests in the USA, CMA has proved to be non-corrosive and considerably less damaging to plants than salt (Horner, 1988). Its major drawback is its cost, which is about 20 times that of salt. Limited research conducted by the author is currently underway to determine the effects of CMA, relative to salt, on species commonly planted in the British roadside environment.

Amelioration

When trees are in an advanced stage of decline, there is little that can be done to save them. However, if salt damage symptoms are noted in the early stages, measures can be taken to minimise damage. The most effective treatment where soil contamination is the primary source of injury is a combination of application of gypsum ($CaSO_4$) to the soil surface (Rubens, 1978), leaching the soil with copious amounts of water and addition of fertilisers low in nitrogen (in the form of nitrate rather than urea; Flückiger and Braun, 1981).

The addition of gypsum to the soil causes Na^+ to be displaced from cation exchange sites in favour of Ca^{2+}. The sulphate (SO_4) in gypsum then combines with the free sodium to form water-soluble Na_2SO_4 which is rapidly leached out of the rooting zone. The removal of Na^+ reduces soil compaction, and the addition of Ca^{2+} helps to restore good soil structure as it facilitates good aeration, percolation, drainage, aggregation and soil capillarity. Increased permeability promotes the leaching of Na^+ and Cl^- ions from the soil.

According to Levitt (1972), Cl^- toxicity is not proportional to the concentration of Cl^- in tissue but to the ratio of the concentrations of Cl^- and SO_4. Chloride toxicity may therefore be reduced by increasing sulphate in the soil. Moreover, both Ca^{2+} and SO_4 inhibit Cl^- uptake by plants, further reducing Cl^- toxicity. Ca^{2+} also promotes higher K^+ : Na^+ ratios and this is beneficial, because it allows roots to remain selectively permeable, and because K^+ is the ion which most often neutralises Cl^-.

Where possible, addition of gypsum should be supplemented by copious watering to facilitate leaching of Na^+ and Cl^- out of the soil. This is a reasonably effective measure even if gypsum is not used. However, both the prior action of salt and soil leaching may result in nutrient imbalance or deficiency; this should be corrected by the use of fertiliser. If appropriate, soil should first be analysed to determine which ions are deficient and fertiliser treatment should be adjusted accordingly. Phosphate and nitrate, in particular, reduce Cl^- uptake and increase exchangeability of Ca^{2+} for Na^+. However, fertilisers with very high concentrations of nitrogen should not be used as this will stimulate growth and further increase uptake of Na^+ and Cl^-. Organic matter increases the effectiveness of gypsum, so addition of fine-textured and well-decomposed mulch is worthwhile for trees of high amenity value.

Application of gypsum to established trees should be on the soil surface of as much of the rooting area as possible and at a rate of approximately 2–3 kg m^{-2}. For new planting, gypsum should be mixed in with good quality backfill as a protective measure. Incorporation of gypsum into the soil is more effective than surface applications. As gypsum is fairly insoluble, a single application may continue to have benefits for 2–3 years.

Obviously, the outlined measures are only practical under certain circumstances. In paved areas, application of gypsum is difficult and lifting of slabs can be very costly. Resources should therefore be directed towards the most accessible trees and trees of high amenity value. Incorporation of gypsum into backfill during planting is probably the most cost-effective measure. Treatment is not recommended if a tree has lost a number of sizeable branches as recovery is unlikely. Treatment will also be of little benefit where other stresses predominate.

Trees should be planted as far back from the road edge as is feasible and the area of open ground around the tree should be as large as possible. Where practical there should be a 'lip' around the edge of the planting pit to divert salty run-off away from the rooting area, or the soil level around the tree should be slightly raised to encourage water to flow away from the tree. In the Netherlands, trees are often protected by straw matting placed around the trunks during the winter, acting as a barrier to salty splash and preventing salty snow being piled at the base of trees.

There is no known ameliorative treatment for salt spray damage. Antidesiccant sprays have been tested as a preventative measure but have proved ineffective (Emmons *et al.*, 1976). Spraying foliage with phosphate may reduce damage, but no trials have been carried out to assess its potential. Where salt spray is a problem, tolerant species should be planted.

Costs: a case study

The Forestry Commission has had links with the arboricultural staff of the City of Westminster for a number of years and research conducted mainly in this Borough first brought recognition of the role of de-icing salt in crown dieback of London plane. This has enabled an assessment of the costs directly attributable to salt damage to be made. P. Akers (Principal Arboricultural Officer, City of Westminster, London) has estimated that de-icing salt was directly responsible for the death of 80 mature planes (out of a population of approximately 5000 plane trees) in the severe winter of 1986/87. The costs in terms of tree removal, replanting and loss of amenity value (calculated according to the Helliwell system; Helliwell, 1967) are shown in Table 18.2.

ACKNOWLEDGEMENTS

This paper is based on a literature review carried out under contract to the Department of the Environment, whose financial assistance is gratefully acknowledged.

Table 18.2 Costs resulting from crown dieback in London plane in the City of Westminster in 1986/87 due to de-icing salt (data from P. Akers).

	Cost (£)
Direct	
Removal of dead trees and roots	300
Planting replacement tree	200
Indirect	
Loss of amenity value	1500
Total	
Per tree	2100
For 80 trees	16800

REFERENCES

AUDIT COMMISSION (1988). *Improving highways maintenance. A management handbook*. HMSO, London.

BERNSTEIN, L., FRANCOIS, L.E. and CLARK, R.A. (1972) Salt tolerance of ornamental shrubs and ground covers. *Journal of the American Society of Horticultural Science* **97**, 550–561.

DAVISON, A.W. (1971). The effects of de-icing salt on roadside verges. I. Soil and plant analysis. *Journal of Applied Ecology* **8**, 555–561.

DEFRAITEUR, M.P. and SCHUMAKER, R. (1988). Plateau des Hautes–Fagnes ou plateau des fontaines-salées? Une nouvelle atteinte à la réserve naturelle: les sels de deneigement. *Hautes Fagnes* **1**, 9–13.

DEPARTMENT OF TRANSPORT (1987). *Winter maintenance of motorways and other trunk roads: statement of service and code of practice*. Department of Transport HM4. DoE/DTp publication.

DIRR, M.A. (1975). Effects of salts and application methods on English ivy. *HortScience* **10**, 182–184.

DOBSON, M.C. (1991). *De-icing salt damage to trees and shrubs*. Forestry Commission Bulletin 101. HMSO, London.

EMMONS, A., WOOD, A. and SUCOFF, E. (1976). *Antidesiccant sprays and damage from deicing salts*. Minnesota Forestry Research Notes No. 258.

FLÜCKIGER, W. and BRAUN, S. (1981). Perspectives of reducing the deleterious effect of de-icing salt upon vegetation. *Plant and Soil* **63**, 527–529.

FOSTER, A.C. and MAUN, M.A. (1978). Effects of highway deicing agents on *Thuja occidentalis* in a greenhouse. *Canadian Journal of Botany* **56**, 2760–2766.

FRENCH, D.W. (1959). Boulevard trees are damaged by salt applied to streets. *Minnesota Farm and Home Scientist* **16**, 9, 22, 23.

GIBBS, J.N. (1983). *Crown damage to London plane*. Arboriculture Research Note 47/83/Path. DoE Arboricultural Advisory and Information Service, Forestry Commission.

GIBBS, J.N. and BURDEKIN, D.A. (1983). De-icing salt and crown damage to London plane. *Arboricultural Journal* **6**, 227–237.

HELLIWELL, D.R. (1967). The amenity value of trees and woodlands. *Arboricultural Journal* **1**, 128–131.

HORNER, R.R. (1988). *Environmental monitoring and evaluation of calcium magnesium acetate* (CMA). Program Report National Cooperative Highway Research No. 305. Transportation Research Board, Washington.

LEVITT, J. (1972). *Responses of plants to environmental stresses*. Academic Press, New York, 489–543.

PARMENTER, B.S. and THORNES, J.E. (1986). *The use of a computer model to predict the formation of ice on road surfaces.* Research Report, Transport and Road Research Laboratory, No. 71. Crowthorne, Berkshire.

RUBENS, J.M. (1978). Soil desalination to counteract maple decline. *Journal of Arboriculture* **4,** 33–43.

SPOTTS, R.A., ALTMAN, J. and STALEY, J.M. (1972). Soil salinity related to ponderosa pine tipburn. *Phytopathology* **62,** 705–708.

SUCOFF, E. (1975). *Effects of deicing salts on woody plants along Minnesota roads.* Technical Bulletin, Minnesota Agricultural Experimental Station No. 303.

Discussion

J. Kopinga ('De Dorschkamp', Wageningen)
The use of urea in de-icing can affect soil conditions by reducing oxygen and increasing uptake of nitrogen.

M. C. Dobson
Agreed. Use of urea is also expensive.

J. N. Gibbs (Forestry Commission)
The use of gypsum to rectify salt damage is still at the experimental stage, though the results are good?

M. C. Dobson
Gypsum has been widely used on coastal reclamation for soil improvement, but there is need for research into its use for ameliorating the effects of de-icing salt.

P. White (London Borough of Ealing)
What time of year should we apply gypsum?

M. C. Dobson
Add with backfill to newly planted trees. Otherwise apply in autumn for either curative or prophylactic purposes.

R. P. Denton (Robert Denton Associates Ltd)
Is there any trend in the increase or decrease in the use of salt?

M. C. Dobson
In Sweden and Germany there is already a reduction in response to reports of pollution and corrosion. In Britain the trend is hopefully decreasing by storing and applying salt with greater efficiency.

Paper 19
Watermark disease of willow

J.G. Turner, K. Guven, K.N. Patrick and J.L.M. Davis, Watermark Research Laboratory, School of Biological Sciences, University of East Anglia, Norwich, NR4 7TJ, U.K.

Summary

By far the most serious disease to affect tree willow is watermark, caused by the bacterial pathogen *Erwinia salicis*. Within the U.K. it poses a serious problem only in East Anglia, principally on cricket bat willow (*Salix alba* var. *caerulea*), but isolated outbreaks have recently been found outside this area. The symptoms and methods of spread are described and the risk to increasingly common amenity willow plantings is emphasised. Diligence is called for from arboriculturists to prevent the spread of this disease in amenity willows.

Introduction

Tree willow in its many species and varieties occupies an important place in the British landscape. The cricket bat willow, *Salix alba* var. *caerulea*, dominates parts of the landscape in Essex and other counties in East Anglia, where it is grown for the commercial production of wood for the cricket bat industry. Elsewhere, willow is grown as an amenity species in landscapes and parks, in urban and suburban developments and it also occurs in wild populations, particularly in low-lying or damp places. As an amenity tree it is valued for the ability to establish readily from unrooted cuttings and to grow rapidly to maturity, producing attractive green foliage. These qualities in particular make willows suitable for use as nursery trees and as screens, for example, between residential areas and business sites.

By far the most serious disease to affect tree willow is watermark, caused by the bacterial pathogen *Erwinia salicis*. Although tree willows are grown throughout Europe, watermark poses a serious problem within the U.K. only in East Anglia, in the Netherlands and, to a lesser extent, in Belgium. Isolated outbreaks have been found recently outside these areas, in Wiltshire and in Leicestershire, and the disease has also been reported in Germany. It has a very patchy geographical distribution and the factors that have defined its present range are likely to include the distribution of extremely susceptible varieties of *S. alba*, the rate of spread of the disease and the effectiveness of local control measures.

Many amenity willow plantations growing outside East Anglia are vulnerable, and likely to be affected by watermark, if the disease becomes more widely established in this country. Because watermark is not generally known to willow growers outside East Anglia attention needs to be drawn to the seriousness of the disease, its symptoms and the control measures that should be applied when these are observed.

Symptoms

The development of symptoms of watermark disease is well documented in *Salix alba*, the white willow, and its varieties (Preece, 1977). The first indication that a tree is infected is in the newly expanded foliage, which withers and then turns a reddish brown. In the early stages, red leaves typically appear in the foliage of one or two branches only; these first signs of disease may easily be missed by inexperienced observers. During the season, affected branches may die back and red leaves occur in adjacent foliage, but after July there is rarely any further advance in these external symptoms, until the following year. In subsequent years, the infected

Plate 19.1 *Dieback symptoms due to watermark in a cricket bat willow.*

Plate 19.2 *Watermark stain in the cut bole of a cricket bat willow.*

branches usually become defoliated, giving rise to a 'stag headed' crown (Plate 19.1). Secondary growth of bushy shoots often develops on branches formerly showing dieback, and the symmetry of the growth of foliage on the crown becomes distorted. The tree rarely recovers from the disease at this stage. It usually becomes progressively disfigured with watermark, growth is reduced and, if the tree is not felled, it may eventually be killed by the disease. It is uncommon to find the disease in trees under 8 years old. It does not develop in the 1-year-old branches of pollarded willows, even though the bole of the tree may be infected (de Kam, 1983).

The most characteristic symptom of watermark is staining of the wood. Cuts made through branches showing the red leaves or dieback usually reveal dark red or brown discoloration of parts of the previous year's annual ring, and the colour darkens when the cut surface is exposed to the air. In young branches, staining may occupy the entire ring, but, in older branches and in the bole, the stain may appear as irregular blotches (Plate 19.2). Occasionally, the inner core of the wood is also stained. If the bark is peeled away, the stain may be observed as streaks in the cambium corresponding to the position of the stain in the underlying wood. The stain is produced by the woody tissue, which is normally white, in response to the activity of *Erwinia salicis*, which grows in the water-conducting vessels of the wood. As a consequence, the water supply to the affected tissue is reduced and this contributes to the development of the red-leaf symptom.

The pathogen is a bacterium <2 μm long, which escapes in vast numbers from the infected wood through fissures, especially in the crotch of branches, to contaminate the outer surfaces of the bark in a slime. These bacteria have been found on the surfaces of leaves and branches of neighbouring trees, possibly as the result of rainsplash dispersal of the exudate (Zweep and de Kam, 1982).

At the University of East Anglia in Norwich, and at the Institute for Forestry and Urban Ecology in Wageningen, the Netherlands, diagnosis of watermark is normally confirmed with the aid of an enzyme-linked immunoassay (ELISA) for *E. salicis*. The assay appears to be specific for *E. salicis*, and we have found no cross-reaction with other species of *Erwinia*, nor with other, unidentified, bacteria commonly isolated from diseased willow wood (Table 19.1). Because this test relies on detection of the pathogen, it can be used to detect infection in symptomless trees, and in species of willow displaying atypical symptoms. For example, diagnosis of watermark in *S. fragilis* is difficult because red leaves usually do not occur, even when the wood shows the characteristic watermark stain (Nash, 1963).

Disease spread

The disease is transmitted over short distances from diseased to healthy trees. Evidence from growers of cricket bat willow over the past 50 years shows that, unless diseased trees are promptly removed, neighbouring healthy trees are likely to become infected. The disease may be spread rapidly between trees planted close together, e.g. <5 m spacing in cricket bat willow. The mechanism of short-distance transmission is not known with certainty but:

surfaces of branches and leaves of healthy trees growing in the vicinity of diseased trees do become contaminated with cells of *E. salicis*;

while the surfaces of willow twigs and branches are contaminated with cells of *E. salicis*, any of several natural physical and biological agents that cause wounds through the bark could introduce the pathogen into the wood;

Table 19.1 Specificity of polyclonal antisera raised against *Erwinia salicis* 'National Collection of Plant Pathogenic Bacteria 2535'.

Bacterium	Number of isolates	Competitive ELISA reaction
Erwinia salicis	57*	+
Erwinia raphontici		−
Erwinia herbicola	1	−
Erwinia chrysanthemi pv. chrysanthemi	1	−
Erwinia chrysanthemi pv. diffenbachiae	1	−
Erwinia amylovora	1	−
Erwinia carotovora var. carotovora	1	−
Enterobacter agglomerans	1	−
Escherichia coli B	1	−
Pseudomonas syringae pv. phaseolicola	4	−
Pseudomonas syringae pv. atropurpurea	1	−
Pseudomonas syringae pv. syringae	1	−
Unidentified contaminating bacteria (not *E. salicis*) isolated from watermarked willow	5	−

* Fifteen isolates were from the National Collection of Plant Pathogenic Bacteria, 17 were from the Netherlands and the remainder were isolated from diseased trees in Essex and Suffolk in 1989.

it is possible to induce watermark disease artificially, with a success rate of approximately 10%, by directly introducing the pathogen into wounds made in the wood with a knife.

This indicates the likely importance of contamination of natural wounds by splash-disseminated bacteria in the short-distance spread of the disease. Another possible source of infection for short-distance spread is the bacterial ooze emanating from the cut surface of stumps remaining after watermarked trees have been felled and removed (Wong *et al.*, 1974).

Watermark disease may also be transmitted in propagating material (Wong *et al.*, 1974; Gremmen and de Kam, 1975). For example, cricket bat willow is grown from 4-year-old cuttings, or 'sets' cut from the mother 'stool', and it has long been known that the disease will develop in sets as they grow on watermarked stools. There is evidence linking the outbreak of some instances of watermark to the planting of diseased sets. For example, a current outbreak of watermark in a single plantation of *S. alba* var. *caerulea* in Wiltshire is the first authenticated record of the disease in that county, and it has developed in trees grown from sets obtained from East Anglia, where the disease is endemic. The most likely source of the infection was in one or a few of the sets, subsequent spread occurring through short-distance transmission between adjacent trees. Paradoxically, it was to escape watermark disease in East Anglia that has led growers to consider establishing cricket bat willow plantations outside this region. This illustrates how the failure to plant disease-free stock or to remove the diseased trees once diagnosed can easily lead to a serious local epidemic.

Susceptibility of willow species

Watermark is primarily a serious disease of *S. alba* and its varieties. Some forms of *S. alba* are

Table 19.2 Species and cultivars of willow susceptible to watermark.

Common name	Scientific name	Cultivar	Country
Watermark recorded			
White willow	*Salix alba* L.	–	England, Netherlands, Belgium
		Liempde*	Netherlands
		Drakenburg*	Netherlands
		Calva	Netherlands
		Belders*	Netherlands
		Lichtenvoorde*	Netherlands
		Tristis	Netherlands
	Hybrids of *S. alba* × *S. fragilis*		
Cricket bat willow*	*S. alba* var. *caerulea*		England
Golden willow	*S. alba* var. *vitellina*		England
Goat willow	*S. caprea*		England
Grey willow	*S. cinerea*		England
Crack willow	*S. fragilis*		England
Purple willow	*S. purpurea*		England
Almond willow	*S. triandra*		England
Watermark induced by artificial inoculation, but no record of natural infection			
Silver willow	*S. alba*	Sericea	England
Bay willow	*S. pentandra*		England
Creeping willow	*S. repens*		England
Osier willow	*S. viminalis*		England

* Extremely susceptible

particularly susceptible, e.g. *Salix alba* 'Liempde' and the cricket bat willow, *S. alba* var. *caerulea*. *E. salicis* also affects most other species of the tree willow, but only infrequently. Watermark has not been reported in the common weeping willow or the *S. babylonica* hybrids. Table 19.2 compiles data from several studies (Wong *et al.*, 1974; Wong and Preece, 1978; Miller-Jones, 1979; de Kam, 1984). Far more attention has been paid to the occurrence of watermark in commercially important willow and in willow grown widely for amenity purposes, than in the other species of willow in which a low incidence of the disease or atypical symptoms may have escaped undetected.

Watermark in cricket bat willow (Salix alba *var.* caerulea)

More than 90% of the wood for the manufacture of the world supply of cricket bats is grown in East Anglia; Essex alone produces one half of this. In the 1930s production of the wood and its associated industry was severely affected by watermark, which, at its height, affected over a quarter of the trees in commercial plantations. The situation was extremely serious, demanding prompt and decisive action to avert a disaster to the cricket bat industry. Acting on evidence that watermarked trees were likely to be an important source of infection for the spread of the disease, Essex County Council obtained statutory powers to appoint officers authorised to enter premises, inspect willows growing there, and to serve notices requiring the destruction of infected trees. The original order, referred to as the Watermark Disease (Essex) Order 1933, was extended over the next 20 years to cover Suffolk, Hertfordshire, Middlesex, Cambridgeshire, Norfolk and Bedfordshire. However, the accurate detection and diagnosis of watermark requires considerable experience, and the field work in other counties has therefore been carried out on an agency basis by officers of Essex County Council. In East Anglia, *ca.* 1.5 million trees are currently in commercial production. The willow inspectors make annual inspections of plantations in Essex containing more than 250 trees;

the smaller plantations are visited less often, on a rotation (Wortley, 1989). Inspection is done during the summer when the red leaf symptom can be seen. During the winter, the inspectors visit the set beds after the sets have been harvested, to examine the cut surfaces of the stools for evidence of the watermark stain.

There can be no doubt that the administration of the Disease Order has been effective in the control of watermark disease in East Anglia. Since the inception of the order in 1933, the incidence of disease has been reduced from *ca.* 25% to the present 0.2%. Further reduction in the incidence may not be possible by this measure alone, but it is generally accepted by the growers that, if inspection was ever relaxed, the incidence of the disease in bat willow plantations would soon return to its former high levels.

At present, the Watermark Disease (Local Authorities) Order (1974) operates only in Essex, Suffolk and Bedfordshire. Other local authorities wishing to adopt these powers should consult the Plant Health Officer, Forestry Commission, 231 Corstorphine Road, Edinburgh, EH12 7AT.

Watermark in the Netherlands

Willows grown for amenity purposes in landscapes and parks and in urban and suburban plantings are also at risk as is clearly shown by the current epidemic in the Netherlands, where it affects, almost exclusively, *S. alba* and its varieties. This example illustrates how easily the epidemic can grow to unmanageable proportions if the disease is not checked in the early stages.

Willows have been grown in the Netherlands for more than 2000 years and, although watermark was first reported in 1932 (though the causative agent was at first wrongly identified; Gremmen and de Kam, 1970), it was probably present much earlier. It may have gone undetected in the pollarded willows grown to provide sticks for firewood and fences, because the symptoms rarely develop in the young branches of these trees, even though the bole may be infected. After 1945 the pollarding of many willows was discontinued and the disease was subsequently found in their maturing branches. These trees are believed to have been the source of infection that spread within rural areas, and to willows in new urban and suburban developments.

During the 1950s, the urban area nearly doubled and there was a large increase in suburban building. Green areas were planned in these developments and were widely planted with fast-growing species, particularly willow and poplar. In the Amsterdam area, 97% of all willow planted as amenity, especially in parks and landscapes, was *S. alba* and, of this, 66% consisted of *S. alba* 'Liempde', highly prized for its attractive foliage. Unfortunately, this cultivar has proved particularly susceptible to watermark and, by the mid-1960s, when the trees were 15–20 years old, it became apparent that the willows were dying on a massive scale. In many cases, afflicted trees died within one growing season.

There was no legislation to enforce the removal of infected trees, but in the urban areas around Amsterdam, willow inspectors were nevertheless appointed and diseased trees were removed. This action was probably too little and too late because watermark in the Amsterdam area continued to increase, reaching a peak in the early 1980s, when records show that 400–600 trees were dying each year. More than 30% of willow trees in the area have now died, or have been removed, as a result of watermark, at a cost to the city authorities of approximately £1 million. To put the problem in perspective, average losses of trees in Amsterdam during 1973–1985 were 240 elm trees per year due to Dutch elm disease (*Ophiostoma ulmi*) (representing an average reduction in the elm population of 0.46% per year) and 268 willow trees per year due to the watermark disease (representing an average reduction in the willow population of 2.1% per year) (Couenberg, 1989).

The same story was repeated in towns and villages across the country. Local efforts to control watermark were probably confounded by the lack of any national policy to fell and destroy diseased willow, especially pollarded willows which are such an integral part of the Dutch landscape, yet remain a source of infection for watermark. There still is no disease-resistant alternative cultivar to the susceptible Liempde and, in the absence of enforced national control, the planting of this very susceptible cultivar has continued to the present day. In fact, in many areas of the Netherlands where the soil is shallow and the water table is high there is no suitable alternative to willow. Therefore the epidemic continues to progress unabated, and the eventual control of watermark in the Netherlands appears to depend upon the future development of disease-resistant cultivars.

There are several possible explanations for the difference between the extensive spread of watermark throughout amenity plantations of willow in the Netherlands, while similar populations in England appear to remain largely unaffected. The watermark symptoms may have simply appeared first in the generally older willow plantations in the Netherlands, as these have approached maturity; latent infections may yet appear in English plantations. Another possibility is that, when the pollarding of willow was largely discontinued in the Netherlands, new willow plantations became exposed to sources of the pathogen more widespread than those occurring in this country, where the disease remains largely confined to East Anglia. It is also possible that the epidemic of watermark in the Netherlands may be caused by a strain of *E. salicis* more virulent than any occurring in England.

Strains of Erwinia salicis

Most plant pathogens exist as a mixture of strains that differ in their aggressiveness, host range, or other characters. Since 1989 we have been examining the strain variation within *E. salicis*, and the geographical distribution of these strains. We did not use host-inoculation tests because of the difficulty of obtaining successful infection under laboratory conditions adequate to contain the pathogen. We have therefore used physiological tests, bacteriophage reaction and immunological tests to distinguish the strains. In a collection of 57 cultures, we have reliably identified five strains that show a distinctive, non-overlapping geographical distribution (Table 19.3). The cultures from the Netherlands formed two strains, D and E, which have not yet been found in diseased trees outside that country; the other strains comprise cultures isolated from England only. Thus, we have been able to confirm, by bacteriophage typing, the previous reports of physiological differences between the Dutch and English strains of *E. salicis*. The urgent need now is to determine whether representatives of the Dutch and English strains also differ in their virulence on willow. We shall also be extending our examination of the strains isolated from watermarked trees in England and will be particularly interested to receive notification of watermark in any of the many trees, grown for landscaping in this country which have been raised from cuttings imported from the Netherlands.

Control of watermark

There is no known resistance to watermark in tree willow but some cultivars are known to be extremely susceptible and should not be used in new plantings of amenity willow (Table 19.3). Cricket bat willow growers do not have this option, of course, because only *S. alba* var. *caerulea* has the particular quality of wood required by bat manufacturers. Propagating material should be purchased from reputable suppliers who check that the mother tree or stool is free from watermark. To identify sources of infection, it is helpful to maintain records of the suppliers of all propagating material.

The site is also important. If watermark occurs in the vicinity, it may be worth planting species other than willow. Spacing the trees so that the branches do not touch at maturity and interspersing willow with other species should reduce the risk of the development of an epidemic if one tree becomes diseased.

As the trees approach maturity, periodic checks should be carried out for signs of watermark, if necessary with the assistance of the Essex County Council Arboricultural Officer. There is evidence that latent infection of watermark may remain dormant for years before the outward symptoms of disease appear. It is important, therefore, to continue periodic inspections even if watermark has not previously been detected in the region.

At the earliest sign of disease, the tree should be felled and destroyed by burning. If practicable, the stump also should be removed and destroyed. Larger stumps may be treated with a suitable herbicide, e.g. glyphosate, which should help to destroy the infection and prevent regrowth. Records indicate a high probability that watermark will develop in willow replants in the vicinity of the stump of the watermarked tree they replace (Wong et al., 1974). Replacement of watermarked trees with species other than willow should therefore be considered.

Table 19.3 Differentiation of strains of *Erwinia salicis*.

Strain	Number of cultures	Reaction to bacteriophage*					Heat-stable antigen†	Galactose use‡	Country of origin
		L	8	3	U	C			
A	36	+	+	–	+	+	+	+	England
B	2	+	–	+	–	–	+		England
C	2	+	+	–	–	+	–	+	England
D	3	+	–	–	–	–	–	–	Netherlands
E	14	+	–	–	–	–	+	–	Netherlands

* Plaque-forming bacteriophage (strains L, 8, 3, U and C) were isolated from soil by enrichment technique, and a drop containing >10^9 particles was placed on a growing lawn of the bacterium. +, lysis; –, no lysis.

† Immunodiffusion tests were used to identify precipitin lines forming between polyclonal antiserum raised against *E. salicis* 'NCPPB 2535' and cells subjected to 121°C for 15 min. +, precipitin line; –, no precipitin line.

‡ Determined as the growth, recorded as optical density (*D*) at 550 nm, in minimal liquid media containing 1% w/v D(+) galactose. +, growth to $D > 0.1$; –, growth to $D < 0.1$.

Conclusion

Watermark is a serious vascular disease of tree willow that has proved particularly difficult to eradicate once it becomes established in plantations of *S. alba* and its cultivars. In commercial plantations of cricket bat willow, watermark is being contained by existing control measures which involve frequent inspection and felling of diseased trees. However, it is important to consider the vulnerability of amenity willow which has been planted in increased number in England since the 1960s. Watermark in willows in Amsterdam did not appear until the trees were 15–20 years old; it is vital that our amenity willow is carefully monitored for signs of the disease so that action can be taken at an early stage to avert further spread.

REFERENCES

COUENBERG, E.A.M. (1989). The significance of *Salix* in urban plantations and the losses caused by the watermark disease in Amsterdam. In *Watermark disease of willow*, ed. M. de Kam, 15–122, De Dorschkamp, Institute for Forestry and Urban Ecology, Wageningen, The Netherlands.

GREMMEN, J. and de KAM, M. (1970). *Erwinia salicis* as the cause of dieback in *Salix alba* in the Netherlands and its identity with *Pseudomonas saliciperda*. *Netherlands Journal of Plant Pathology* **76**, 249–252.

GREMMEN, J. and de KAM, M. (1975). The necessity of using healthy propagating material of *Salix alba* in connection with the spread of watermark disease in the Netherlands. *European Journal of Plant Pathology* **51**, 376–383.

de KAM, M. (1983). Watermark is not transmitted with one-year-old cuttings of *Salix alba*. *European Journal of Forestry Pathology* **12**, 365–376.

de KAM, M. (1984). Het vastellen van de gevoeligheid van wilgen voor de watermerkziekte: problemen en perspectiven *Nederlands Bosbouwtijdschr* **56**, 22–27 (Dutch, with English summary).

MILLER-JONES, D.N. (1979). *Watermark disease in wild and cultivated willows*. Ph.D. thesis, University of Leeds (unpublished).

NASH, T.H. (1963). *Watermark disease,* Erwinia salicis *bacterium, scourge of the cricket bat willow*. Land Agent and Valuers Department, Essex County Council, Chelmsford.

PREECE, T.F. (1977). *Watermark disease of the cricket bat willow*. Forestry Commission Leaflet 20. HMSO, London.

WONG, W.C., NASH, T.H. and PREECE, T.F. (1974). A field survey of watermark disease of cricket bat willow in Essex and observations on some of the probable sources of the disease. *Plant Pathology* **23**, 25–29.

WONG, W.C. and PREECE, T.F. (1978). Infection of cricket bat willow (*Salix alba* var. *caerulea* Sm.) by *Erwinia salicis* (Day) Chester detected in the field by use of specific antiserum. *Plant Pathology* **22**, 95–97.

WORTLEY, M. (1989). Legislation as a means of controlling *Erwinia salicis* in Great Britain. In *Watermark disease of willow*, ed. M. de Kam, 67–70. Institute for Forestry and Urban Ecology, Wageningen, The Netherlands.

ZWEEP, P. van der and de KAM, M. (1982). The occurrence of *Erwinia salicis*, the cause of the watermark disease, in the phyllosphere of *Salix alba*. *European Journal of Forestry Pathology* **12**, 257–261.

Discussion

N. Fay (Treework Services Ltd)
Are there any variations in the disease due to seasonal variations, i.e. drought, and is the virulence related to times of cutting?

J.G. Turner
No to both questions. The data are only related to cricket bat willows. The disease takes a long time to develop and a long time to emerge. The Dutch believe there is an increased risk of infection when water tension is high. We have found it difficult to get infection from inoculations and have not proved this increased risk.

W.E. Matthews (Southern Tree Surgeons Ltd)
Is this not like a rabies situation, where we should ban imports of willow from the Netherlands?

J.G. Turner
Imports of willows from the Netherlands are from regularly inspected suppliers. The difference in virulence of the Dutch strain of the disease is unknown and therefore difficult to ban.

Paper 20

Recognition and investigation of unexplained disorders of trees

R.G. Strouts, Forestry Commission Research Station, Alice Holt Lodge, Farnham, Surrey, GU10 4LH, U.K.

Summary

Over the years, Forestry Commission pathologists have become familiar with a number of distinctive, yet unexplained, disorders of ornamental trees, and new ones continue to come to light. In this paper, these are grouped according to the degree of difficulty each presents to the investigator, from those which are evidently due to an infectious organism and for which a number of possible candidates have been identified to those which appear to be unlike any described in the literature and for which no likely cause is evident.

Introduction

The vast majority of problems reported to the Forestry Commission's Pathology Diagnostic and Advisory Service are caused by well-known disorders, pests or diseases, such as herbicide damage, aphid defoliation or honey fungus (*Armillaria*). At the end of each year, however, a number of intractable puzzles remain, not because of lack of information or inappropriate specimens, or because the tree has died and the cause long-since vanished, but because the problems have not been solved by the routine investigations. These remain unsolved, perhaps for years, despite repeated efforts to solve them. Such a 'perennial puzzle' is distinctive enough to be recognisable as the same syndrome time and time again.

Such occurrence may be frequent in certain years, as in the recent *Prunus* 'Kanzan' mystery (Plate 20.1) (Strouts, 1991); or it may come to our notice only once or twice over a long period of years, perhaps on several trees in each case. Almost invariably the problem crops up on a particular tree species or even on a particular cultivar (as in the 'Kanzan' puzzle; indeed, this is one reason for assuming that the problem is a repetition of a previously unexplained one). The symptoms may suggest that the cause is a living organism – perennial 'target' cankers are almost invariably caused by a fungus or bacterium, for example; or one or more organisms may have been isolated from the dying tissue; or there may be some other clue as to the cause – for example, the onset of the trouble might follow particular weather conditions, or cases may be confined to one geographical area or site type.

Unexplained diseases with an associated organism

Potentially, the easiest type to investigate is where one, or several, named organisms are regularly associated with the damage, none of which is known to cause such damage, and yet the damage is typical of the type caused by an infectious organism. In an uncommon, but very distinctive disease of yew, odd branches, large or small, die, clearly (on anatomical examination) because patches of bark, more extensive each year, die and either girdle the branches or spread proximally, killing the branches as they go. The one fungus isolated from many such branches is a *Phomopsis* species, but no *Phomopsis* disease of yew, or any similar disease of yew, is described in the literature, and inoculations with this species into healthy yews on several occasions have always failed to result in the development of the disease. The next step

161

would be to vary inoculation techniques or to inoculate at different times of year and to look for and test other organisms if new cases should come to light.

Investigations into a similar, very common and distinctive branch-killing disease of cherry laurel, *Prunus laurocerasus* (Plate 20.2), have reached this stage on several occasions. A *Phomopsis* species and the fungus *Cytospora* had been isolated from dying bark but neither was a known pathogen. After several attempts at different times of year, inoculations with *C. laurocerasi* did induce the disease.

Unexplained diseases associated with an unidentified organism

More of a problem are those instances, such as yew branch dieback, where an obviously infectious disease is involved and a fungus has been isolated, and yet not only is the disease apparently undescribed but the fungus cannot be identified. Without a name, it is impossible to check the literature to see whether the fungus is known to be pathogenic. It is then necessary to test the fungus on healthy trees and to continue efforts to identify it. For a disease of scarlet oak (Plate 20.3) which fell into this category, inoculations were dramatically successful but whether this was a hitherto undescribed disease was still uncertain as the fungus was still unidentified. Eventually an identifiable fungal fruit body was found on dying tissue, and an isolate from this proved to be none other than the common *Bulgaria inquinans* (Plate 20.4).

Mr Bill Matthews of Southern Tree Surgeons, a well-known member of the Arboricultural Association, pointed this disease out to my predecessor in respect of scarlet oaks in Sheffield Park in Sussex in 1964 and Mr Matthews himself suggested that *Bulgaria* was the cause. It is nice to be able to confirm his suspicions now, albeit 26 years late! Our doubts at that time hinged on the inconclusive results of the small amount of published work which dealt with the possible pathogenicity of *Bulgaria*, on the unusual species of tree involved, and on the striking similarity of symptoms of this disease to a disease caused by a *Phytophthora* species. Only after attempts to find *Phytophthora* in diseased bark repeatedly failed did it become clear that other lines of enquiry should be pursued.

Unexplained diseases and disorders with no associated organism apparent

It is just as difficult to further investigations into a case which has all the hallmarks of a particular disease or disorder but where this diagnosis, based on symptoms and circumstances, cannot be confirmed. Eventually such problems may be resolved, but several are currently still mysterious. For example, Lombardy poplars with partially dead crowns (Plate 20.5) are not hard to find, but the cause is. Everything suggest that this is the fungal disease *Dothichiza*, but no one has been able to find this fungus on dying parts, nor has any other satisfactory explanation for the condition been put forward.

Sometimes, though, these seemingly intractable problems are resolved. A disease of wild cherry, *Prunus avium* (Plate 20.6) characterised by the death of branches which exude a conspicuous brown gum, which also occurs on other flowering cherries, has been familiar for many years. Its symptoms fit so precisely the descriptions given for that plague of cherry and plum orchards bacterial canker (*Pseudomonas syringae* pv. *mors-prunorum*) that we have confidently and repeatedly diagnosed it as such. Not until last year was this diagnosis confirmed. Following a visit to a site with us, a bacteriologist from East Malling (Horticultural Research International) isolated the bacterium from *Prunus avium* trees showing exactly these symptoms. In previous years, enquiries had been received and our isolations made in summer, which is too late in the year for the bacterium to be isolated from the bark lesions.

Sometimes the symptoms and circumstances of these puzzling cases suggest an insect problem. For example, a disorder of cedar in Kew Gardens had all the hallmarks of a heavy infestation of a sucking insect: browning and falling needles, stunted growth and thick deposits of sooty moulds, presumably growing on honey

Plate 20.1 Unexplained dieback of Prunus 'Kanzan'. Circumstantial evidence suggests this may be caused by the bacterium Pseudomonas syringae *pv.* syringae. (39016)

Plate 20.2 A recently explained branch and stem-killing disease of cherry laurel. The fungus Cytospora laurocerasi *girdles branches and causes a true dieback.* (35805)

Plate 20.3 Result of infecting the bark of scarlet oak artificially with the fungus Bulgaria inquinans. (R. G. Strouts)

Plate 20.4 Fruit bodies of Bulgaria inquinans. (37463)

Opposite

Plate 20.5 *Unexplained branch-killing disease of Lombardy poplar. The symptoms closely resemble those attributed to the fungus* Dothichiza (Discosporium) populea *in Europe and North America.* (39254)

Plate 20.6 *Severe dieback of wild cherry* (Prunus avium) *caused by the bark-killing bacterium* Pseudomonas syringae *pv.* mors-prunorum. (36908)

Plate 20.7 *Yellowing and loss of all but the current year's needles and death of some one-year-old shoots of yew, due to drought.* (R. G. Strouts)

Plate 20.8 *An unexplained, fatal disease of* Crataegus oxycanthoides *(red-flowered cultivar) characterised by black flecks in the outer few annual xylem rings of the rootstock (probably* C. monogyna*). Note that there are no flecks above the graft line.* (39258)

dew. Careful examination by entomologists failed to reveal any other sign of aphids or other sucking insects, nor was the damage familiar enough for them to diagnose it as caused by any particular pest. It was not until the following year that this puzzle was resolved, when our entomologists found the culprits: the aphid *Cedrobium laportei*. At about the same time, an entomologist from the British Museum, whose daily journey to and from work took him through Kew Gardens, found this same insect on the damaged Kew trees. As with some bacteria and fungi, insect pests are often only present at certain times of the year and have often vanished by the time symptoms have drawn attention to the problem.

These kinds of puzzles, where no causal agent is present, are even more difficult if suspicion falls on some non-living agent; drought, frost or lightning cannot be induced to reveal themselves on diseased tissues in the laboratory. Periodically, older needles on yew turn yellow and fall (Plate 20.7). No sign of a living cause has been found on such trees, so the possibility of weather-induced damage must be considered. As the damage on several occasions has quickly followed periods of drought and as similar damage on other conifers is undoubtedly a drought effect, drought is probably the cause of the damage on yew also. Unless experimentation is possible, such diagnoses, made on the basis of circumstantial evidence, remain valid as long as a more convincing explanation is lacking, or until there is contrary circumstantial evidence.

Unexplained and previously unrecorded diseases

Rarely, a 'perennial puzzle' can be clearly characterised, yet nothing like it is reported in the literature and no likely cause comes to mind. These are the hardest of all to deal with. One such problem which has exercised us over the last 2 years is a strange and fatal condition of hawthorn (all the few cases so far have been on red-flowered cultivars of *Crataegus oxycanthoides*). The tree dies, probably as a result of death of the roots. The striking feature is that the rootstock (presumably *Crataegus monogyna*) exhibits small but numerous and prominent blackish flecks (Plate 20.8) in the outer few rings of the wood; these do not extend into the scion (the *C. oxycanthoides* part of the tree), and the graft line is clearly visible where stained and healthy wood meet. Although *C. monogyna* is so common, we have not yet seen this flecking in grafted or ungrafted individuals of this species, though so few cases of the disease have come to light that this may merely be chance. Cultures from these flecks have produced no organism sufficiently often to suggest a causal agent. On the freshest example of such a tree that we have examined, dying roots and other symptoms of *Phytophthora* root disease were present and *Phytophthora* was isolated from these. Whether this fungal pathogen is involved in the black fleck disease or whether this was the chance occurrence of two diseases on the same tree is still unclear.

Few of the difficulties described have proved insurmountable and gradually 'perennial puzzles' are solved, but, to the irritation of tree owners and to the benefit of plant pathologists, new puzzles continue to arise.

REFERENCE

STROUTS, R.G. (1991). *Dieback of flowering cherry*, Prunus 'Kanzan'. Arboriculture Research Note 94/91/PAT. DoE Arboricultural Advisory and Information Service, Forestry Commission.

Discussion

R.G. Pawsey (Pathologist and Consultant)

Can you tell us something about the handling of queries following reorganisation of the Pathology Section at Alice Holt?

R.G. Strouts

There is now limited time to deal with enquiries. There will, therefore, be a reduction in time spent on enquiries relating to privately owned non-woodland trees and more time spent on research. More queries will be passed to the Arboricultural Advisory and Information Service.

T. Walsh (Birmingham City Council)
What advances have been made on slime fluxing and disruption of bark on red horse chestnut?

R. G. Strouts
Phytophthora cactorum and bacterial wetwood is found on horse chestnut. The cause of bark eruptions of red horse chestnut is not known.

W. E. Matthews (Southern Tree Surgeons Ltd)
Browning has occurred on evergreens a long way inland this winter. Is this salt damage?

R. G. Strouts
It is likely to be salt in the wind. Chloride analysis of foliage should confirm this diagnosis.

M. C. Dobson (Forestry Commission)
Foliage analysed at Alice Holt Lodge from trees some 30 miles from the sea showed higher chloride concentrations than normal although there was no strong browning of the leaves.

Paper 21
Recent advances in detection of wood decay

D. A. Seaby, *Plant Pathology Research Division, Department of Agriculture, Newforge Lane, Belfast, BT9 5PX, U.K.*

Summary

A wide range of methods used to detect wood decay is reviewed and two novel devices made by the author are described for detecting decay (or severe insect damage) within beams or standing trees.

The portable compression meter consists of a stiff narrow probe with a slightly enlarged tip which is driven progressively into a pre-drilled 4 mm diameter hole. It is driven forward by an engineer's automatic hand punch which provides uniform pulses, which are counted for each centimetre of penetration. The number of pulses cm^{-1} decreased significantly on reaching decayed timber.

The other device consists of an electric hand drill with a modification for maintaining constant drill-bit pressure combined with a capacity to record drill-bit rotation cm^{-1} of penetration. Within the gravimetric density range 0 – 0.67, drill-bit rotations cm^{-1} of penetration (r/p) were highly correlated with the cube of the density, and decay was detected by a sudden reduction in r/p.

Examples of how these devices have been used are described and their potential usefulness is discussed.

Introduction

The ideal method of detecting and quantifying timber decay in the field should be rapid, portable, inexpensive and non-destructive. Ultrasonic, sonic and X-ray detectors may eventually meet most of these criteria, but are currently costly and bulky. X rays may be ruled out in some circumstances because they have to pass through the test material; they are unlikely ever to be user friendly (Habermehl, 1982a, b). The same applies to γ radiation from a radioactive source (Tiler, 1972). Nevertheless, computed tomography (X-ray scanning of a log to reconstruct its internal structure from multiple projections) shows up such features as insect attack, geometry of a defective core, and bark and resin pockets. In these images, variations in green-wood density are also shown and can be measured by image analysis (Benson Cooper *et al.*, 1982). Some advances in X-ray computer tomography are described by Funt and Bryant (1987). X-ray densitometry is now an established method for assessing whole-stem density for use in some tree improvement programmes (Gonzalez, 1987). Sonic and ultrasonic detectors (McCracken and Vann, 1983; Dunlop, 1981) may eventually be more practical for field work but will still require skilled interpretation of results for quantification of decay or computer analysis after scanning from various angles (Bulleit and Falk, 1985; Beal and Wilcox, 1987).

Line (1984) found catalase enzyme activity a useful non-destructive adjunct to visual assessment of stakes or wood blocks.

In Sweden, sniffer dogs are used to detect wood rot (M. Lothian, personal communication); in Canada, magnetic resonance imaging using a whole-body NMR scanner has been used to detect the hidden morphology of wood (Hall *et al.*, 1986).

A number of semi-destructive techniques are available. All these create a borehole in the test piece and this may enhance conditions for further decay development.

The Pressler and French auger removes cores of wood which can be visually examined by eye or under the microscope. These cores may be used for a limited number of small-scale mechanical tests for hardness etc. (P. Savill, personal communication) or, in the case of incipient decay, for immunological tests to detect the presence of wood-decaying fungi (Houghton et al., 1987). Alternatively, if extracts are made from samples of the wood using hot water, infra-red spectrophotometry can be used to detect early stages of brown rot which are associated with an infra-red absorption peak at 1720 cm^{-1} (Gibson et al., 1985). Wood decay can also be detected by the use of scanning calorimetry because wood-rotting fungi cause changes in the relative sizes of the cellulose and lignin peaks (Reh et al., 1986). A simpler method using colour saturation after staining with bromophenol blue allowed objective identification of spruce wood decay (Katuscak and Katuscakova, 1987).

The more sensitive the method the smaller the sample that needs to be taken. This is desirable because the major disadvantages of corers is that they leave a large test hole, up to 1 cm in diameter. They are also tedious to use, even when motorised. Blair and Driver (1977) described the use of a light-weight, petrol-powered drill to obtain wood chips which were then cultured to detect wood-rotting fungi.

British Telecom detect rot in telegraph poles using a rod with an enlarged threaded tip. This is manually screwed into the test pole. Rot is detected by a reduction in resistance (T. Boyce, personal communication). Methods of detection of rot in Swedish utility poles are reviewed by Henningsson (1985).

An ingenious indirect method of detecting decay is the well-publicised shigometer (Shigo and Berry 1975; Gallagher and Syndor, 1983; Wilkes and Heather, 1983; Wilson, 1983; Shigo, 1984). An electric probe is slowly inserted into a narrow pre-drilled hole and the pattern of resistance of the wood to a pulsed current is recorded. This method depends on the concentration of cations being increased by the activity of wood-decaying fungi. The results, which can be confusing, also depend on the tightness of fit of the electrode and the moisture versus resin content of the timber (Wilson et al., 1982).

In future, gas chromatography may be used to detect gaseous fungal metabolites in wood after extracting a headspace sample from a borehole. The method may, however, be restricted to active rot produced by specific fungi and is unlikely to give quantifiable results.

Two simple direct methods for the quantification of decay are now described.

Portable compression strength meter (PCM)

The PCM consists of a stiff graduated probe up to 28 cm long with an enlarged rounded tip. The probe fits on to an engineer's automatic marking punch, which has a graduated impact adjustment. The punch transforms a slowly increasing pressure from the hand into a sudden reproducible impact which drives the probe forward along a preformed 4 mm diameter drill hole. As the probe tip progresses it enlarges the drill hole by crushing the side walls. The energy to do this per centimetre of travel is proportional to the number of punch blows required.

The device was tested by Barrett et al. (1987) using a log of mature pine wood decayed internally by *Phaeolus schweinitzii*. Seven test lines were made through the log and from these the volume of decay was mapped and then compared with a visual assessment after dissection. The PCM accurately detected both the visible rot and the incipient rot (wood in which mycelium was subsequently discovered). The PCM also detected a zone of harder reaction wood around the rotted area. This 'reaction zone' (*sensu* Shain, 1967) was also detected in a test on a birch log (see Table 21.1).

Analysis by Barrett et al. (1987) of the data after comparison between PCM values and values provided by standard strength tests demonstrated a highly significant linear relationship between gravimetric density of the wood and PCM strikes ($r^2 = 0.8$) and between compression to failure point and PCM ($r^2 = 0.81$). In a further test I found that, within a range of five softwoods and seven hardwoods, the correlation between PCM strikes and density remained high ($r^2 = 0.78$).

Table 21.1 Portable compression meter (PCM) values over 16 cm radius of an internally decayed birch log. The decayed area of 10–16 cm is 65% weaker than the outer wood. There is a dense reaction zone at 8 cm in from the bark. (Barrett, 1987)

Radial distance from bark (cm):	1	2	3	4	5	6	7	8	9	10	11	12	13	14	15	16
PCM values	12	10	14	15	15	15	13	24	16	8	5	4	4	3	4	5

In a field test on living trees, an unexpected relationship between the wood hardness of certain species and their compass aspect was noted. Six trees of each of five species were chosen at random at the centre of three blocks of trees in Belvoir Forest, Co. Antrim, Northern Ireland. These were tested with a PCM at breast height on their south, east, north and west aspects. Four species, *Pinus contorta*, *Pinus sylvestris*, *Abies grandis* and *Picea abies*, showed significantly greater PCM values to the east with minimum values to the west or south; *Larix decidua* showed little variation (Figure 21.1). No trends relating to planting lines were noted. It is considered likely that the growth rate to the west and south was fractionally greater than to the east or north and that this accounted for the observed variations. This phenomenon is reported to highlight the potential sensitivity of the PCM and to show that compass aspect may have to be taken into account when comparing the compression strength of individual standing trees.

Details of the PCM system are available from Barrett (1987).

Decay-detecting drill (DDD)

Two versions of this drill exist: a battery-powered field prototype (Figure 21.2), which records drill-bit rotations as a series of dots on graph paper, and a mains-powered prototype which records drill rotations per centimetre of penetration, on a series of electronic counters. The field version is built round a two-speed reversible drill. An adaptation to the drill was made to provide a second chuck mounted on a telescopic spring-loaded tube. Pressure on the drill bit compressed this spring and, at a predetermined pressure position, a small light-emit-

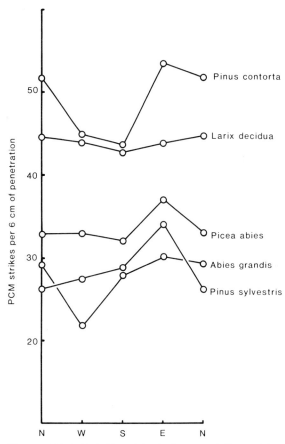

Figure 21.1 *Relationship between strikes with a portable compression strength meter (PCM) and compass aspect for five forest tree species.*

ting diode was activated. This device enabled constant drill bit pressure to be applied by the operator. The fine drill bit, >20 cm long and 1.7 mm in diameter with a flared 2.0 mm diameter cutting tip, was supported by a 3-cm long bearing attached to a bracket at the end of a telescopic carriage, on which the recording

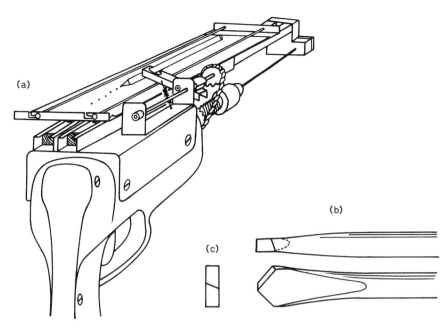

Figure 21.2 *Field version of decay-detecting drill (a); enlarged side view (b) and end on view (c), illustrating drill bit design.*

paper was mounted. In use, the carriage was pushed back along a trackway as the drill bit penetrated the test piece. This moved the graph paper past a mounted pen which, as a result of the action of a cam driven by a gear chain from the chuck, made a visible dot on the graph paper after each 2.5 rotations of the drill bit. In hardwood, the dots were very close together because the bit penetrated slowly, but in softwood the dots were widely spaced. Rot was detected by a sudden widening in the spacing of the dots.

The DDD was extremely sensitive, recording the pattern of hardwood grain, knotty areas and reaction wood as well as decayed wood. Some examples of traces from healthy wood are shown in Figure 21.3a.

To investigate the relationship between DDD traces and standard wood tests, three replicate 4-cm long test borings were made in 19 wood samples ranging from balsa wood to oak.

When the number of revolutions cm^{-1} of penetration (r/p) was plotted against wood density, a curvilinear relationship emerged. This was transformed to a linear relationship in the density range 0–0.67 by plotting r/p against density[3] (Figure 21.4). Deviation from the linear relationship depended mainly on the variable moisture content, which particularly affected the density of softwoods. Reproducibility for particular samples of seasoned timber was very high, with standard errors for three replicates averaging <2% of the mean.

In standing trees, the sap and resin content resulted in progressively greater drag on the drill bit with depth of penetration. Rotational drag did not affect readings but longitudinal drag reduced drill-bit tip pressure by c. 10% cm^{-1}. This consequently produced results which gave the impression that the wood density steadily increased with depth of penetration whereas in dry stem discs the reverse was usually true. This drawback still allowed some density comparison between trees by comparing traces of equal length. Drag was reduced by fitting an oil-saturated lubrication pad just in front of the drill-bit support bearing. Decay was easily detected as a sudden reduction in r/p (see Figure 21.3b).

In a recent modification of the drill, a cylinder was made to rotate in synchronisation with the drill bit. A sheet of paper was wound around

171

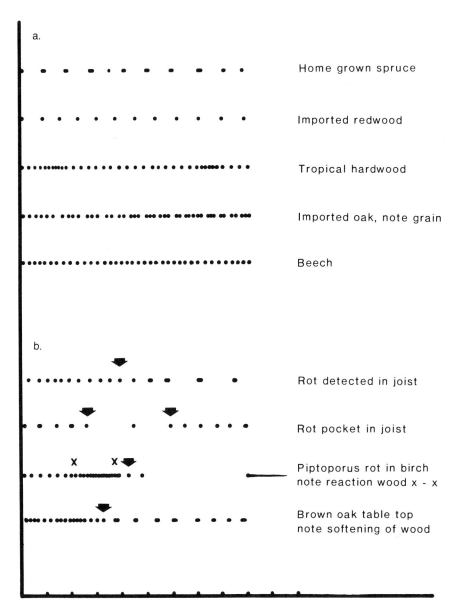

Figure 21.3 *Parts of traces made by decay-detecting drill in (a) healthy wood of five tree species and (b) decaying wood.*

this cylinder and then a spring-loaded pen was made to impinge on it. As the drill bit penetrated the test piece, the pen moved down the rotating cylinder, thus producing a spiral trace (Figure 21.5). The trace for pine represented very even close-grained wood (Figure 21.5a) with approximately eight annual rings cm^{-1}. In spruce (Figure 20.5b), the trace was produced by

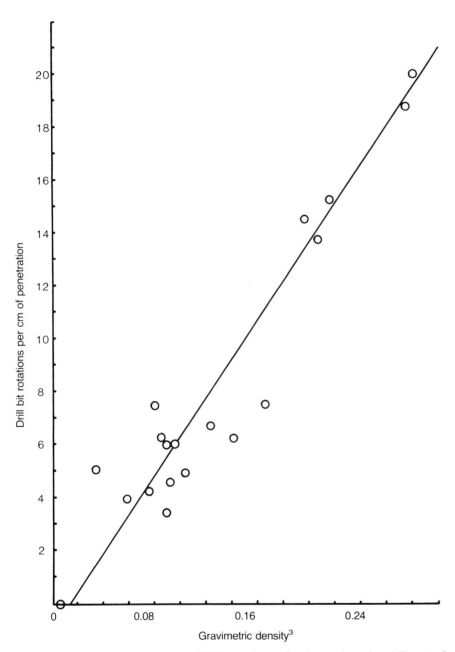

Figure 21.4 *Relationship between drill-bit rotations of a decay-detecting drill cm^{-1} of penetration and gravimetric density^3 for 19 wood samples, ranging from balsa wood to oak, correlation 0.929.*

drilling into uneven-grained wood with less than two annual rings cm^{-1}; two of these annual rings were resin soaked and showed up clearly as dense horizontal lines on the trace. Bands of spring wood were detected as steeply sloping sections in the spruce trace. With other samples, rot and woodworm tunnels showed up as near-vertical lines.

Discussion

Uses of the PCM and DDD are complementary.

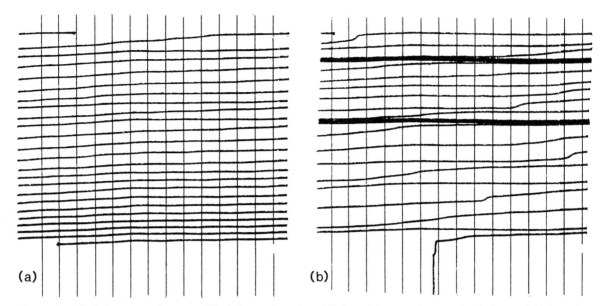

Figure 21.5 *Spiral traces, using a modified decay-detecting drill, from (a) even, close-grained pine and (b) irregular, coarse-grained spruce; the thick lines show two resin-soaked rings.*

The PCM is precise but much slower to use than the DDD. Although the PCM is easily carried in a pocket it requires a drill to create the lead hole. The PCM appears much less affected by the moisture and resin content of the standing tree so that it would be the preferred choice for detailed intertree comparison for the assessment of timber quality or for elite tree selection. However, it requires a 4 mm diameter hole, which, although much smaller than that made by the Pressler borer, is still over twice the diameter of the DDD bit which creates almost negligible damage.

Design development of the drill-bit tip may reduce drag in standing timber and further improve drill performance. Potential forest uses for the DDD include quantifying decay in stands of timber to be sold at auction. This is likely to be an important use because currently prices are sometimes significantly depressed by uncertainty of the quantity of rot present in a stand, bids erring on the side of caution. Aboricultural uses would be mainly the detection of hidden rot in standing trees.

A major use for the DDD is foreseen in the detection of decay when surveying houses. Beams and joists, even when covered by carpet and floor boards, can be drilled into and tested without doing appreciable damage. Ship's timbers, railway sleepers and wood transmission poles could also be easily tested, thus improving safety standards.

The DDD is the subject of a patent application.

REFERENCES

BARRETT, D.K., SEABY, D.A., GOURLAY, I.D. (1987). Portable compression strength meter; a tool for the detection and quantification of decay in trees. *Arboricultural Journal* **11**, 313–322.

BARRETT, D.K. (1987). *Users guide to the PCM system*. Obtainable from Wood End, 54 Upper Road, Kennington, Oxford OX1 5LJ.

BEAL, F.C. and WILCOX, W.W. (1987). Relationship of acoustic emission during radial compression to mass loss from decay. *Forest Products Journal* **37,** (4) 38–42.

BENSON COOPER, D.M., KNOWLES, R.L., THOMSON, F.J. and COWN, D.J. (1982). *Computed tomographic scanning for the detection of defects within logs*. Bulletin of the New Zealand Forest Research Institute No. 8.

BLAIR, H.A. and DRIVER, C.H. (1977). A portable power drill can aid the detection of tree root and butt rot pathogens. *Plant Disease Reporter* **61**, 481–483.

BULLEIT, W.M. and FALK, R.H. (1985). Modelling of stress wave passage times in wood utility poles. *Wood Science and Technology* **19**, 183–191.

DUNLOP, J.I. (1981). Testing of poles using the acoustic pulse method. *Wood Science and Technology* **15**, 301–310.

FUNT, B.V. and BRYANT, E.C. (1987). Detection of internal log defects by automatic interpretation of computer tomography images. *Forest Products Journal* **37**, 56–62.

GALLAGHER, P.W. and SYNDOR, T.D. (1983). Electrical resistance related to volume of discoloured and decayed wood in silver maple. *Horticultural Science* **18**, 762–764.

GIBSON, D.G., KRAHMER, R.L. and DE GOOT, R. C. (1985). Early detection of brown rot decay in Douglas fir and southern yellow pine by infra-red spectrophotometry. *Wood and Fibre Science* **17**, 522–528.

GONZALEZ, J.S. (1987). Relationship between breast height and whole stem density in 50 year old coastal Douglas fir. In *Proceedings of the 21st meeting of the Canadian Tree Improvement Association*, part 2. Truro, Nova Scotia.

HABERMEHL, A. (1982a). A new non-destructive method for determining internal wood conditions and decay in living trees. I. Principles, method and apparatus. *Arboricultural Journal* **6**, 1–8.

HABERMEHL, A. (1982b). A new non-destructive method for determining internal wood condition and decay in living trees. II Results and further developments. *Arboricultural Journal* **6**, 121–130.

HALL, L. D., RAJANAYAGAM, V., STEWART, W. A., STEINER, P. R. and CHOW, B. (1986). Detection of hidden morphology of wood by magnetic resonance imaging. *Canadian Journal of Forest Research* **16**, 684–687.

HENNINGSSON, B. (1985). *The preservative treated utility pole in service: research and experience in Sweden*. Uppsater, Institutionen for Virkeslara, Sveriges Lantbuksuniversitet, No. 150.

HOUGHTON, D.R., SMITH and EGGINS, H.O.W. (1987). Immunological methods for the detection and characterisation of wood decaying basidiomycetes. *Biodeterioration* 7. Selected papers presented at the 7th International Biodeterioration Symposium Cambridge, UK, Sept 1987.

KATUSCAK, S. and KATUSCAKOVA, G. (1987). Means of objective identification of spruce wood decay. *Holzforschung* **41**, 315–320.

LINE, M.A. (1984). The potential utilization of catalase detection techniques for the non-destructive assay of wood decay. *International Biodeterioration* **20**, 85–91.

McCRACKEN, F.I. and VANN, S.R. (1983). *Sound can detect decay in standing hardwood trees*. USDA Forest Service, Southern Forest Experiment Station, Research Paper, SO–195.

REH, U., KRAEPELIN, G. and LAMPRECT, I. (1986). Use of scanning calorimetry for structural analysis of fungally degraded wood. *Applied and Environmental Microbiology* **52**, 1101–1106.

SHAIN, L. (1967). Resistance of sapwood in the stems of loblolly pine to infection by *Fomes annosus*. *Phytopathology* **57**, 1034–1045.

SHIGO, A.L. (1984). How to assess the defect status of a stand. *Northern Journal of Applied Forestry* **1**, 41–49.

SHIGO, A.L. and BERRY, P. (1975). A new tool for detecting decay associated with *Fomes annosus* in *Pinus resinosa*. *Plant Disease Reporter* **59**, 739–742.

TILER, T.H. (1972). Gamma radiation detects defects in trees and logs. *Phytopathology* **62**, 756.

WILKES, J. and HEATHER, W.A. (1983). Correlation of resistance to pulsed current with several wood properties in living eucalypts. *New Zealand Journal of Forestry Science* **13**, 139–145.

WILSON, P. J. (1983). The shigometer technique in practice. *Arboricultural Journal* **7**, 81–85.

WILSON, P.J., ALLEN, J.D. and WALKER, J.C.F. (1982). Appraisal of Shigometer technique. *New Zealand Journal of Forestry Science* **12,** 86–95.

Discussion

M.P. Denne (Bangor University)
 The density of timber varies within and between species, so there is need for knowledge about natural variations between species.

D.A. Seaby
 The method of density detection described uses sudden differences to indicate decay, e.g. unpredictable softness in heartwood.

M.P. Denne
 What about the difference in density between summer and spring wood of conifers?

D.A. Seaby
 This occurs but does not give as sudden a change in readings as between sound wood and decayed wood.

B.J.W. Greig (Forestry Commission)
 Will the decay detection drill be available commercially?

D.A. Seaby
 A prototype has recently been completed by P. and J. Sibert, Pyrford, Woking, Surrey. An initial production of six drills will be made and tests carried out by British Telecom and Rentokil.

Paper 22

Tree decay in relation to pruning practice and wound treatment: a progress report

D. Lonsdale, Forestry Commission, Alice Holt Lodge, Farnham, Surrey, GU10 4LH, U.K.

Summary

Progress of research is reported in biological control of decay, evaluation of a novel and durable wound sealant and analysis of factors in wound creation which influence the development of decay.

In a 5-year (1984–1989) trial of the biocontrol agent *Trichoderma* isolate 127 on fourteen tree species, the final data from four species showed an absence of basidiomycete colonisation and the persistence of *Trichoderma* in 83–100% of wounds. For eight of the remaining species, similar data were obtained in a one-year interim sampling, but Lombardy poplar (*Populus nigra* 'Italica') showed a high incidence of basidiomycete colonisation after 3 years (1984–1987). A physical wound sealant with high durability reduced microbial colonisation beneath beech wounds in a one-year pilot trial more than any material previously tested.

Wound closure on beech and *Sorbus intermedia* was improved in the adaxial and abaxial positions by the retention of the branch collar or branch ridge, as compared with flush pruning. In an experiment involving nine tree species, the season of pruning had highly significant effects on the initial rate and pattern of wound closure, cambial dieback and staining of wood beneath wounds.

Introduction

At the last in this series of seminars, we considered a number of advances in research and in understanding which could perhaps help us to develop ways of lessening the risk of serious decay developing beneath pruning wounds (Shigo and Marx, 1977). For example, it was then clear, at least for beech (Mercer, 1982), that the extent of decay tends to increase with wound size. There was growing acceptance that the risk is also increased by the use of flush cuts, as opposed to those which retain the branch collar (if such is visible) or the 'branch bark ridge' (Shigo, 1982) As far as wound treatments were concerned, no chemical or physical dressing had been found capable of preventing microbial colonisation, despite the inclusion of a wide range of such materials in studies in the U.K. (Mercer *et al.*, 1983) and the USA (e.g. Shigo and Shortle, 1983).

Although there is now a wide awareness of new approaches to tree pruning, the advice offered to practitioners has not taken the form of specific guidelines, except in the case of the 'ridge' pruning technique. For example, there has been advice to avoid making excessively large wounds wherever feasible, but the advisable size limit has not been defined with any certainty and indeed may be expected to vary according to factors such as tree species, the number of wounds per tree and the time of year of cutting. Similarly, the guidance on wound treatments has been rather negative; i.e. untreated wounds can be invaded by decay fungi, but treatment is not worthwhile except in special circumstances, for example, when fresh-wound parasites are likely to be a problem (Dye and Wheeler, 1968; Wilkes *et al.*, 1983; Clifford and Gendle, 1987). On the other hand, data obtained by Mercer and Kirk (1984), who evaluated the biological control agent *Trichoderma* sp. on beech stem wounds, had indicated by the early 1980s that there was a basis for promoting this form of treatment. A need remained for further evaluation of *Trichoderma* on pruning wounds of a range of tree species.

Some of these remaining questions have been investigated within the present project over the last 5 years and the following lines of enquiry have now generated sufficient data for a progress report to be made:

i. the evaluation of the biocontrol agent *Trichoderma* isolate 127 on pruning wounds of fourteen tree species;
ii. preliminary evaluation of a novel wound sealant which has been selected for its exceptional durability and hence its potential to act like a natural wound occlusion in inhibiting the growth of decay fungi;
iii. experimental evaluation of creating wounds at 'stub', 'flush' and 'ridge' positions; and
iv. the influence of the season of pruning on cambial dieback, cicatrix (botanical term used here to describe the new bark and wood formed around a wound; the term 'callus' is commonly used to describe this, but more properly refers to an undifferentiated growth of cells) formation and staining beneath the cut wood surface, as observed in nine tree species and on wounds either treated with a sealant or untreated.

Materials and methods

In the experimental work reported at the last seminar (Lonsdale, 1987), the pruning wounds created were all of the 'stub' type. Sampling of the stubs allowed efficient evaluation of a wide range of wound treatments, but it was important in subsequent work to carry out an assessment of the changes occurring within the xylem after the creation of more typical pruning wounds closer to the branch base. The 'ridge' method of pruning was therefore adopted, despite the need for wounds created by this method to be sampled by dissecting or boring the parent stem.

Except in the evaluation of pruning wound position, the experimental procedure for wounding experiments was as follows:

a. selection of a tree from which at least one set of replicate branches (usually three or four per set) could be cut without reducing crown volume by more than 20%;
b. selection of suitable, unsuppressed branches which had basal diameter above flare at least 50 mm (usual range 50–85 mm), no cankers or dieback near base and, wherever possible, no ingrown bark between branch and parent stem;
c. creation of a pruning wound at the 'ridge' position, following preliminary cuts to remove the weight of the branch, and examination to ensure the absence of any existing staining or decay in the wood;
d. recording of wound dimensions, local diameter of the parent stem and height above ground;
e. immediate application of any treatment;
f. incubation period (from a few days to 5 years);
g. measurement of cicatrix formation at the margin of the wound, and of cambial dieback in certain experiments;
h. exposure of internal tissues for sampling, usually by dissection of the parent stem, but in some cases by removal of an increment borer core; and
i. recording of data on: depth and spatial pattern of staining or decay, depth of penetration by each fungal or bacterial species present and identification of micro-organisms where appropriate (e.g. in the case of basidiomycetes or introduced biocontrol agents).

Experimental details and results

Evaluation of the biocontrol agent *Trichoderma*, isolate 127

A 5-year trial was set up over 1984–1987 on beech, pedunculate oak, wild cherry, silver birch, Lombardy poplar, Norway maple, sycamore, ash, common lime, hornbeam, holly, yew, sallow and horse chestnut. The biocontrol agent was applied to the freshly cut wound surfaces as a spore suspension (10^8 spores ml^{-1} of water) and a supplementary covering of the proprietary sealant Lac Balsam was added as soon

Table 22.1 Preliminary data from trial of biocontrol of decay with *Trichoderma* (isolate 127) on trees

Tree species	Percentage of wounds tested			
	Isolation of Trichoderma		Isolation of basidiomycetes	
	Sealant* only	Trichoderma + sealant	Sealant* only	Trichoderma + sealant
Sampling after 5 years†				
Fagus sylvatica	78	83	0	0
Quercus robur	89	94	0	0
Prunus avium	90	100	0	0
Betula pendula	100	100	8‡	8‡
Sampling after 3 years				
Populus 'Italica'	60	90	5	47
Sampling after 1 year†				
Acer platanoides	83	100	0	0
A. pseudoplatanus	30	100	0	0
Tilia vulgaris	100	88	0	0
Carpinus betulus	100	100	0	0
Ilex aquifolium	77	80	0	0
Taxus baccata	75	100	0	0
Salix caprea	100	100	0	0
Aesculus hippocastanum	100	60	0	0

* Lac Balsam applied as an antidesiccant.
† Ten wounds per tree species were included.
‡ Birch wounds containing basidiomycetes contained stained wood before the experiment was set up.

as the suspension had been absorbed. Control wounds received Lac Balsam only.

In March 1987, the final sampling on Lombardy poplar was carried out after only 3 years, since many of the wounds were becoming totally occluded because of the extreme vigour of this cultivar. In May 1989, final sampling after 5 years was carried out on beech, birch, wild cherry and pedunculate oak. Final sampling for the other nine species is not due until 1992 but preliminary data were obtained for each of them by plating out wood chips taken from wound surfaces one year after pruning. The data in Table 22.1 show the percentage of wounds from which *Trichoderma* and basidiomycetes (decay fungi) were isolated in the final and preliminary samplings.

Trichoderma was isolated from a high percentage of wounds, even for the four species subjected to final sampling after 5 years. There was also a high incidence of *Trichoderma* colonisation in the 'control' wounds, indicating either that the natural incidence of the fungus in uninoculated wounds was high, or that cross-infection had somehow occurred, despite precautions against it. This high incidence (compared with a typical value of c. 15% of wounds in other experiments and surveys) suggested that cross-infection was more likely than frequent natural infection, and this interpretation was reinforced by the results of 'challenge' tests on agar (Plate 22.1) showing that most of the control isolates were indistinguishable from the inoculated strain.

The data also show that basidiomycete infection was generally absent from wounds treated with

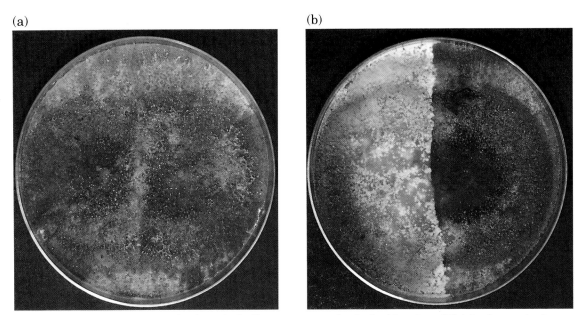

Plate 22.1 *Demonstration of cultural challenge tests to check the identity of* Trichoderma *isolates recovered from pruning wounds in* Trichoderma *biocontrol trials: (a) strain 127 challenged against itself, note the merging of colonies, (b) strain 127 challenged against another strain, note the sharp division between colonies.*

Trichoderma, except in Lombardy poplar, in which *Chondrostereum purpureum*, the silver-leaf pathogen, was found; indeed, on poplar, *Trichoderma* inoculation gave a higher isolation frequency for *C. purpureum*. The basidiomycetes in birch were not fresh colonists, having been unavoidably present in the main stem before the wounds were created. Basidiomycetes were generally absent from 'control' wounds, but this was probably due to the presence of *Trichoderma* from cross-contamination; on the basis of other experiments and surveys, an average of 12% of tree wounds would have been expected to yield basidiomycetes.

Evaluation of a durable physical sealant

Among the materials considered for this purpose on the basis of durability of strength and elasticity was polyurethane in the form of a monomer which polymerised on contact with moisture. In laboratory tests of sealant strength (Lonsdale, 1987), a polyurethane mastic product gave promising results, but its rapid rate of polymerisation made it unsuitable for application to wound surfaces. A new polyurethane-

Table 22.2. Penetration of staining and fungal growth in Isoflex-treated wounds on beech, 12 months after treatment (data based on mean values from 35 mm-deep wood blocks).

	Depth of staining (mm)	Depth of fungal colonisation (mm)	Percentage of samples yielding basidiomycetes
Treated*	6.9	0.5	0
Control†	25.9	18.6	2

* Mean values from 36 wood blocks.
† Mean values from 49 blocks from separate trees.

Table 22.3 Mean closure of pruning wounds (mm) at different positions on beech in the first growing season after pruning; cicatrix formation.

Wound position:	Flush		Ridge		Stub	
Closure direction:	Vertical	Horizontal	Vertical	Horizontal	Vertical	Horizontal
Treatment						
Seal and Heal	3.5	10.6	7.1	5.3	4.9	3.7
Control	1.7	4.8	1.9	3.7	1.7	0.5

based product, 'Isoflex', later became available for use in situations where slow curing was required, principally as a repair compound for felt roofs. This product performed extremely well in laboratory tests and was therefore selected for inclusion in field trials.

A one-year field trial of 'Isoflex' was set up on beech pruning wounds in 1986. The data (Table 22.2) showed only a very slight penetration of staining and microbial colonisation. On the basis of these encouraging results, a longer-term trial was set up in 1989, using beech and oak pruning wounds, for sampling in 1991 and 1994.

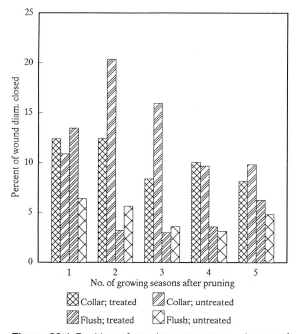

Figure 22.1 *Position of pruning: percentage closure of axial (vertical) collar and flush wounds on* Sorbus intermedia *over 5 years, with or without treatment with Lac Balsam.*

Pruning practices

Effects of creating pruning wounds at 'stub', 'flush' and 'ridge' positions

Experiments involving these three pruning positions and the use of a sealant, 'Seal and Heal' have been set up at three locations, using beech and pedunculate oak in replicated trials, as well as *Torreya californica, Tilia tomentosa, Quercus castanifolia, Pinus radiata* and *Eucalyptus dalrympleana*, added as part of a pilot trial. An experiment on a single specimen of *Sorbus intermedia* had been set up in 1981 to compare flush pruning with collar retention, and this was assessed annually until 1986. The experiments on beech and oak are continuing, but preliminary data for cicatrix formation around beech wounds have been collected at one of the sites.

The data from beech (Table 22.3) show that, after one year, cicatrix formation was generally better at the sides of the wounds than above or below, and that this positional difference was much more pronounced on flush wounds than on ridge or stub wounds. Cicatrix formation on stub wounds was poor at all positions. The ridge wounds gave the greatest cicatrix formation in the vertical direction, and so had a much greater tendency to occlude in a circular pattern, (i.e. all the way round the wound) rather than in a 'slit' pattern. Despite the rapidity of lateral cicatrix formation on flush wounds, ridge wounds tended to become more rapidly occluded since, for a given branch diameter, flush wounds were initially much larger. Similar results were obtained from the *Sorbus* experiment (Figure 22.1), indicating that, where branch collars are visible, their retention in pruning promotes a

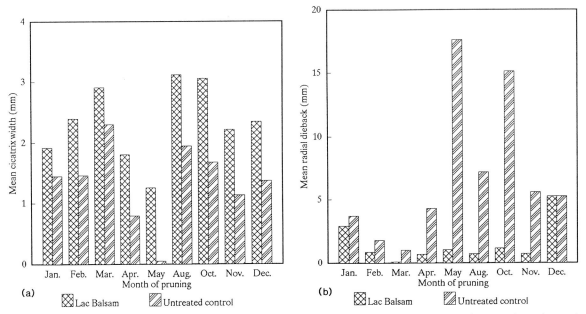

Figure 22.2 Season of pruning: (a) cicatrix formation and (b) cambial dieback around wounds on pedunculate oak after pruning at nine times of year. Values shown are the means of measurements at 12 positions around each wound. Significance: month $P<0.001$; sealant $P<0.001$; month $\times$ sealant not significant in (a), $P<0.001$ in (b).

more satisfactory pattern of occlusion than flush pruning. These data were not, however, statistically significant, probably because of lack of adequate replication.

Effects of creating wounds at different times of year

Beech, wild cherry, sycamore, Norway maple, hornbeam, ash, pedunculate oak, red oak and yew were used in this experiment. The wounds were either left untreated or treated with Lac Balsam. Ten replicates of each combination of species, time and treatment were laid down. In addition to the assessments described under Materials and methods, the temperature was recorded at the time of wounding, in case freezing periods, with possible tissue injury, occurred at any of the times of wounding. Any notable features of the branch base structure, such as fluting of the parent stem, were recorded to detect their relationship, if any, with patterns of dieback and cicatrix formation.

These wounds were assessed after one year and involved four main measurements: (a) the extent of cambial dieback, (b) the extent of cicatrix formation, (c) the depth of staining in the wood beneath the wound surface and (d) the incidence of colonisation by basidiomycete fungi. Cicatrix formation and dieback were assessed using a purpose-made device which allowed measurements at twelve circumferential positions.

The differences between times of year with respect to cambial dieback and cicatrix formation were highly significant for all nine species. The very extensive data sets cannot be published in this paper, but examples are shown here (Figure 22.2), while Figure 22.3 shows a summary of the times of year which are asociated with relatively little cambial dieback and wood staining. For some species, autumn emerged as a rather unsuitable time for pruning, but this was not so for all of them.

Positional differences around the wound circumference were also marked, as shown for pedunculate oak in Figure 22.4, and significant for all the species. Dieback was greatest underneath wounds, and least at the top, while cicatrix formation was greatest at the sides and intermediate in value above the wounds (Plate

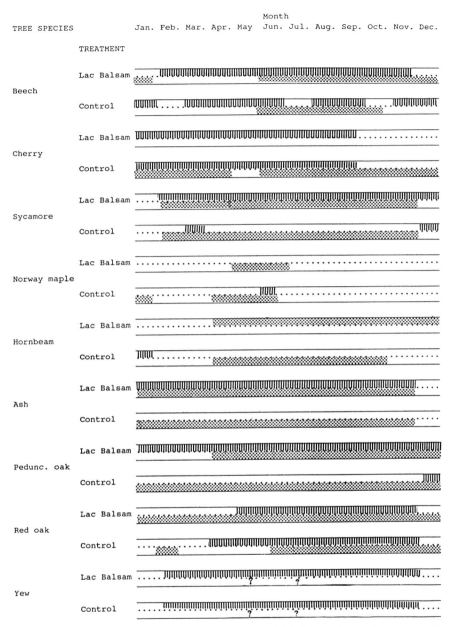

Figure 22.3 Provisional 'diary' of periods when cambial dieback (striped) at any position on the edge of the wound was <10 mm (or <25% of the worst mean value recorded in the control wounds, whichever is the greater) and wood staining (stippled) was <25 mm below the wound surface, in nine tree species. No staining data available for yew.

22.2). Important interactions occurred between season and wound position.

In some cases, treatment of the wounds with Lac Balsam had a significant effect on cicatrix formation or dieback, but the effects were not consistently beneficial, as shown by the summary in Figure 22.3.

Seasonal variations in the depth of xylem staining beneath wounds were also statistically significant and in most species there was a dis-

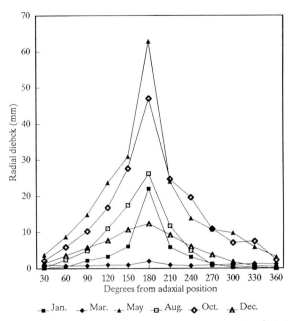

Figure 22.4 Season of pruning: patterns of cambial dieback around wound circumference on pedunculate oak after pruning without treatment with a sealant (for clarity, data for February, April and November are not shown). Significance: month, position around wound and month × position all $P<0.001$.

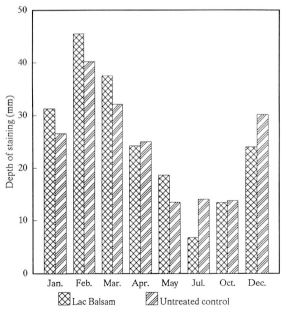

Figure 22.5 Season of pruning: staining of wood beneath wounds made at eight times of year in hornbeam; note the cyclic pattern. Significance: month $P<0.001$; sealant and month × sealant not significant.

Plate 22.2 Seasonal pruning experiment: assessment of dieback around wound (darkened area) on Prunus avium.

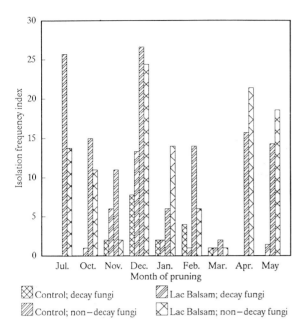

Figure 22.6 *Season of pruning: isolation of fungi from wounds on wild cherry from pruning at nine times of year and treated with or without Lac Balsam. The term 'decay fungi' refers to basidiomycetes.*

tinct cyclic pattern, with a January or February maximum value and a summer minimum (see example in Figure 22.5). Again, treatment with Lac Balsam produced significant effects in some cases.

The frequencies of occurrence of basidiomycetes (decay fungi) and other non-basidiomycete fungi were also recorded, and it was found that the former were not present in wounds made at certain times of year on beech, wild cherry, hornbeam, ash, sycamore and Norway maple. Ash and the two *Acer* species yielded basidiomycetes mainly from wounds created in autumn and early winter, whilst wild cherry yielded these fungi from autumn to spring, with a fairly pronounced peak in December. An example is shown in Figure 22.6.

Discussion

Wound treatments

As far as physical sealants are concerned, the data presented here relate only to 12-month experiments and therefore provide no grounds for questioning the now widely held belief that wound paints do not prevent decay in the long term. Nevertheless, it is interesting that the polyurethane-based 'Isoflex', a sealant which appears to have much better durability than those previously tested, greatly reduced the depth of staining and fungal colonisation of beech wounds, compared with controls. The 5-year trial of this material on beech and oak will indicate whether it can have such an effect in the longer term.

It is also interesting that the proprietary sealant 'Lac Balsam' had significant effects on the extent of wood staining in the seasonal pruning experiment. The sealant decreased staining for many combinations of species and season, but increased it for others. One possibility is that the moisture content of the injured wood can be affected by sealants either to the advantage or disadvantage of invading micro-organisms, depending on its underlying status as determined by tree species and the time of year. Earlier work on stem wounds has shown that sealants can reduce the extent of staining over at least a few years (Rohmeder, 1953; Houston, 1971; Bonnemann 1979), but Shigo and Shortle (1983) have argued that this effect disappears within 7 years. Their argument implies that the migration of the boundary between sound and discoloured wood is, at best, merely slowed down by treatment. It would be interesting to see whether this is true of pruning wounds, in which the boundary seems sometimes to migrate very little from its initial position.

Trichoderma 127 has been shown to establish itself very easily in wounds on a wide range of tree species and to persist better than might have been expected on the basis of earlier work in North America (Pottle *et al.*, 1977); for at least 5 years in four tree species which have been subjected to final sampling. The absence of basidiomycete colonisation of the treated wounds in species other than Lombardy poplar was also encouraging. The success of the treatment can be measured against the fact that basidiomycetes have been found in an average of 12% of pruning wounds in all other experiments and surveys, even though they were absent from control wounds in this experiment, the probable consequence of cross-contamination by *Trichoderma*.

Pruning practices

The data on the position of wounding must be regarded as preliminary in the sense that only wound closure, and not the extent of staining or decay in the wood, has so far been assessed. Nevertheless, the data support the idea that the retention of the ridge or collar is preferable to flush cutting.

The investigation of the effects of pruning at different times of year has shown some important effects with respect to cambial dieback, cicatrix formation and staining. The data are represented here in the form of a calendar of times when cambial dieback or xylem staining fell below certain arbitrary limits. This must be regarded as an example of what could be achieved with further research, since several questions remain unanswered. Important amongst these is whether the results can be repeated, but the sheer scale of the work and the availability of suitable trees will not readily allow a full repetition. However, the work has been repeated on beech and pedunculate oak, at a weekly frequency over the periods when changes seem most rapid: in autumn and spring. In this context it is of course important to take the physiological state of the tree into account, as well as the date. In these studies the question of 'physiological' season is being addressed to some extent by the measurement of electrical resistance in the bark, which alters during spring and autumn.

When this experiment was set up, no comparable work had been found in the literature, but a rather different study involving bark removal on beech, red oak and pedunculate oak has recently been reported by Liese and Dujesiefken (1989), working in Germany. They found that all three species showed better wound closure following wounding in April than in October, December or February.

ACKNOWLEDGEMENTS

I gratefully acknowledge the assistance of Mr I. T. Hickman, Mrs S. E. Brown and Mr C. Palmer who have worked on this project. Permission for the use of trees in experimental work was kindly provided by New Ideal Homes and by the Hillier Arboretum. I am also grateful to the suppliers of 'Isoflex', Sterling Roncraft Ltd. and the former Isoflex Ltd., for providing samples of this product. The project is financed, under contract, by the Department of the Environment.

REFERENCES

BONNEMANN, I. (1979). *Untersuchungen über die Entstehung und Verhütung von 'Wundfaülen' bei der Fichte.* (Unpublished) Doctoral Thesis, Forstlichen Fakultät der Georg–August-Universität zu Göttingen.

CLIFFORD, D.R. and GENDLE, P. (1987). Treatments of fresh wound parasites and of cankers. In *Advances in practical arboriculture*, ed. D. Patch, 145–148, Forestry Commission Bulletin 65. HMSO, London.

DYE, M.H. and WHEELER, P.J. (1968). Wound dressings for the prevention of silver-leaf in fruit trees caused by *Stereum purpureum* (Pers.) Fr. *New Zealand Journal of Agricultural Research* **11**, 874–882.

HOUSTON, D.R. (1971). Discoloration and decay in red maple and yellow birch; reduction through wound treatment. *Forest Science* **17**, 402–406.

LIESE, W. and DUJESIEFKEN, D. (1989). Aspekte und Befunde zur Sanierungszeit in der Baumpflege. *Das Gartenamt* **38**, 356–360.

LONSDALE, D. (1987). Prospects for long-term protection against decay in trees. In *Advances in practical arboriculture*, ed. D. Patch, 149–155, Forestry Commission Bulletin 65. HMSO, London.

MERCER, P.C. (1982). Tree wounds and their treatment. *Arboricultural Journal* **6**, 131–137.

MERCER, P.C. and KIRK, S.A. (1984). Biological treatments for the control of decay in tree wounds. II. Field tests. *Annals of Applied Biology* **104**, 221–229.

MERCER, P.C., KIRK, S.A., GENDLE, P. and CLIFFORD, D. R. (1983). Chemical treatments for control of decay in pruning

wounds. *Annals of Applied Biology* **102**, 435–453.
POTTLE, H.W., SHIGO, A.L. and BLANCHARD, R.O. (1977). Biological control of wound hymenomycetes by *Trichoderma viride*. *Plant Disease Reporter* **61**, 687–690.
ROHMEDER, E. (1953). Wundschutz an verletzten Fichten. *Forstwissenschaftlichen Zentralblatt* **72**, 321–325.
SHIGO, A.L. (1982). A pictorial primer for proper pruning. *Forest Notes* (Society for Protection of New Hampshire Forests), Spring Issue.
SHIGO, A.L. and MARX, H.G. (1977). *Compartmentalization of decay in trees.* USDA Forest Service, Agriculture Information Bulletin 405. 72 pp.
SHIGO, A.L. and SHORTLE, W.C. (1983). Wound dressings: results of studies over thirteen years. *Journal of Arboriculture* **9**, 317–329.
WILKES, T.J., VOLLE, D. and LEE, T.C. (1983). Effect of fungicides on infection of apricot and cherry pruning wounds with *Chondrostereum purpureum*. *Australian Journal of Agriculture and Animal Husbandry* **23**, 91–94.

Discussion

E. Guillot (Prospect Tree Services)
Would cherry trees be damaged if pruned between January and June?

D. Lonsdale
Our results for wild cherry suggest that the amounts of wood staining and cambial dieback will be small if you avoid pruning during October to December and April to May. Your 'windows of opportunity' are narrower than this if you also consider seasonal risks from various diseases to which *Prunus* species are prone. Silver-leaf is one of the worst of these and is best avoided by pruning only during June to August.

R. Harris (Cambridge City Council)
Do experiments have built-in recognition of natural processes, e.g. leaf flush?

D. Lonsdale
It's true that natural processes are a more valid guide than the date, but resources haven't yet allowed us to monitor them. Our follow-up studies will involve weekly pruning intervals, and we will then record phenological stage and cambial electrical resistance.

J. Boyd (Merrist Wood Agricultural College)
Can *Trichoderma* species be suspended in chainsaw oil and then automatically applied to the wound?

D. A. Seaby
There are problems in keeping spores alive in the oil. A suspension lasting several months can be made by mixing lecithin and glycole. Add this to oil and it stays in suspension for a few hours and will therefore be transferred to the cut surface. It has been used on pine stumps (50%) and spruce (100%). Pines are more selective and more prone to *Fomes* infection. A lot of fungal infection occurs at low temperatures, but *Trichoderma* likes higher temperatures.

D. Lonsdale
We have set up trials using 'cocktails' of *Trichoderma* isolates to see which ones take at whatever temperature.

G. Litchfield (Svensk TradVard 85)
Is there a relationship between ambient air temperature and time of pruning, e.g. should cutting be avoided at temperatures below –10°C.

D. Lonsdale
Daytime temperature of –10°C very rarely, if ever, occur in the UK. Frost at pruning time is considered damaging, but in our experiments it occurred too rarely for assessment.

E. Barrs (National Trust)
With regard to *Trichoderma*, what success has there been with different isolates?

D. Lonsdale
We had promising results with a commercial mixture of species, 'Binab T', but this was only a pilot trial.

Paper 23

Stump fumigation as a control measure against honey fungus

B.J.W. Greig, *Forestry Commission Research Station, Alice Holt Lodge, Farnham, Surrey, GU10 4LH, U.K.*

Summary

Fumigants were applied to stumps *in situ* which were infected with honey fungus. The treatments were tested on seven conifer species. Stumps were excavated at different periods after treatment and the extent of *Armillaria* was measured by cultural isolations onto selective media in the laboratory.

Metham-sodium (Sistan) significantly reduced the incidence of *Armillaria* in stump and root samples for all species, even up to 3½ years after treatment. Dazomet (Basamid) also gave a significant reduction but was less effective than metham-sodium. Methyl bromide was only tested in one experiment, where it almost totally eliminated *Armillaria* from Scots pine stumps. There was a high incidence of *Trichoderma* sp. isolated from the treated stumps.

The results show that fumigants have considerable potential for reducing *Armillaria* inoculum in situations where stumps cannot be physically removed from the ground.

Introduction

Honey fungus (*Armillaria* spp.) needs little introduction to most arboriculturists. Its importance in Britain is shown by information provided by the Pathology Department of the Royal Horticultural Society Gardens at Wisley, where at least 400 enquiries about the disease are received each year (P. Greenwood, personal communication). At Alice Holt, from 1984 to 1988 the Pathology Advisory Service dealt with 70 cases each year; these comprising about 10% of all enquiries (R. G. Strouts and D. R. Rose, personal communication). Recommended control measures include the removal of stumps and dead roots and the establishment of barriers to disease spread by trenching or physical means (Greig and Strouts, 1983). In many cases these measures are impractical and there is clearly merit in the development of a treatment which would eliminate, or greatly reduce, honey fungus in stumps and roots *in situ*.

In 1974, Filip and Roth in the U.S.A. tested various soil fumigants for their ability to eradicate honey fungus from stumps of ponderosa pine (*Pinus ponderosa* Laws.; see Table 23.1). The fumigated stumps were excavated after one year and isolations showed that all the fumigants reduced *A. mellea* in the treated stumps as compared with the controls (Filip and Roth 1977). In a similar study, Thies and Nelson (1982) eliminated *Phellinus weirii*, a serious root pathogen in north-west U.S.A. and Canada, from Douglas fir stumps using four fumigants (Table 23.1). Several other workers have used

Table 23.1 Fumigants used to eradicate fungi on tree stumps in trials in the USA.

For *Armillaria mellea* on ponderosa pine*
    Methyl bromide
    Vorlex
    Chloropicrin
    Carbon disulfide
    Vapam

For *Phellinus werii* on Douglas fir†
    Chloropicrin
    Allyl alcohol
    Vapam
    Vorlex

*Filip and Roth, 1977. †Thies and Nelson, 1982

Table 23.2 Details of tree species and *Armillaria* species in trials of fumigants on tree stumps in the UK in 1983–1986.

Experiment number	Tree species	Type of stump	Armillaria species present
1	Lawson cypress	Thinning stumps 18 months old	*A. bulbosa* and *A. mellea*
2	Scots pine		*A. bulbosa* and *A. mellea*
3	Scots pine	Stumps of recently killed trees	*A. ostoyae*
4	Scots pine		*A. ostoyae*
5	Scots pine		*A. ostoyae*
6	Scots pine / Lawson cypress	Thinning stumps 12 months old	*A. bulbosa* and *A. mellea*
7	Serbian spruce / Grand fir / Japanese fir / Douglas fir / Western hemlock		*A. bulbosa* and *A. mellea*

fumigants to eradicate pathogenic fungi from wood blocks buried in soil (Godfrey, 1936; Bliss, 1951; Rackham *et al.*, 1968; Houston and Eno, 1969) and decay fungi from transmission poles (Graham, 1975; Graham and Corden, 1980). Some fumigants were also used successfully to eradicate decay fungi in impregnated Douglas fir timber (Eslyn and Highley, 1985). After reviewing these trials it was decided to test some of these fumigants against honey fungus in Britain; this paper reports the results of trials established in 1983–1986.

Materials and methods

For the U.K. trials, metham-sodium (product name Sistan or Vapam) was chosen because it performed consistently well in the U.S.A. and is safer to use than the other fumigants; it is favoured by U.S.A. scientists for these reasons (G. M. Filip, personal communication). Dazomet (product name Basamid) was not tested in the U.S.A., but like metham-sodium it decomposes in the soil to form methyl isothiocyanate and, as it is in a powder form, it is easy to handle. Methyl bromide was included in one trial, as it generally gave the best results in the U.S.A. but, because it is highly toxic and can only be used in the U.K. by licensed operators, it is unlikely to find widespread application here.

In the experiments (Table 23.2), three of the five British *Armillaria* species (Gregory, 1989), *A. bulbosa*, *A. mellea* and *A. ostoyae*, were present. Two sites were used in the trials. Experiments 1, 2, 6 and 7 were in Alice Holt Forest in Hampshire in a mixed crop of conifers planted in 1961 after the clear-felling of a 140-year-old oak (*Quercus robur*) crop. Both *A. bulbosa* and *A. mellea* occurred on this site. Experiments 3, 4 and 5 were in Warren Wood on the Elveden Estate in Suffolk in a crop of Scots pine (*Pinus sylvestris*) planted in 1951 which had suffered very severely from fatal attacks by *A. ostoyae*. The treated and control stumps were checked for *Armillaria* infection, either by isolation into pure culture or by the presence of the characteristic mycelium beneath the bark.

Vertical holes were bored into the surface of the stump using a 2.5 cm auger in a drill attachment on a chainsaw. On an average size stump of 20 cm diameter, four holes were drilled to a

Plate 23.1 Scots pine stump with holes drilled ready for treatment.

Plate 23.2 Scots pine stump shown in Plate 23.1 after treatment, with holes sealed with hardwood dowels.

depth of 10–30 cm, depending on species. Larger stumps had more holes drilled in them (Plate 23.1). Holes were drilled to the maximum possible depth, without penetration into the soil; if this inadvertently occurred, another hole was drilled. The fumigants were poured into the holes to within 2 cm of the top and these were then sealed using hardwood dowels (Plate 23.2). The fumigants were also applied to the soil at four points around the stumps in holes made by a crowbar, these holes then being sealed with soil. Finally, the stumps were covered for 1–3 weeks with clear polythene sheeting buried under soil at the edges. For the methyl bromide treatment in Experiment 3, the stumps were drilled and the surrounding vegetation was removed. Black polythene sheeting, buried at the edge, was then used to provide a complete sealant over the stump and surrounding area. Methyl bromide was applied at 4 lb 100 ft^{-2} of surface area from cans via tubing which pierced the sheeting over the stumps. After 24 h the sheeting was removed. The control stumps in all experiments were drilled, plugged and covered with polythene, but received no further treatment. In Experiment 4 the stumps were chipped to about 10 cm below ground level, using a portable stump chipper and Sistan was applied to holes drilled into the residue of the stump.

The stumps were excavated using hand-operated Tirfor or Lugall winches. After excavation, the soil was removed and all the roots were sawn off as close to the stump as possible. The stump 'body' was then split into four quarters, and five chips of wood were removed from the exposed faces of each quarter section (Figure 23.1), and placed on a selective medium (Kuhlman, 1966). Discs approximately 2 cm thick were cut at intervals of 15 cm along each root to a root diameter of 1 cm and four isolations were made from the cut surface of each disc. An additional isolation was made from any mycelium or rhizomorph present beneath the bark (Figure 23.2). On average, 75 isolations were made from each stump. Throughout, the isolations were not made from random positions, but were biased towards those areas which appeared to be decayed by *Armillaria*. In examining the plates, records were made not only of *Armillaria* but also of *Trichoderma*, a common genus of the Fungi Imperfecti, which has received much attention over the years as a potential biological control agent for various plant pathogens.

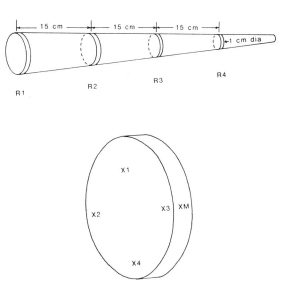

Figure 23.2 *Diagram of a root showing discs cut at intervals of 15 cm and the isolation points (X1–X4 and XM) from one of the discs.*

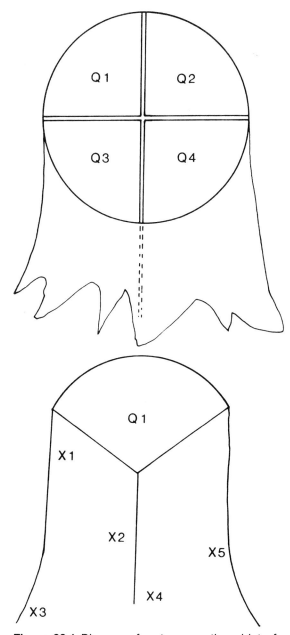

Figure 23.1 *Diagram of a stump sectioned into four quarters (Q1–Q4) and the five isolation points (X1–X5) on one of the quarter sections.*

Results

In the first three experiments, six stumps from each treatment were excavated after 2 years and a further six after 3 years. In Experiment 1 on Lawson cypress thinning stumps, *Armillaria* was isolated after 2 years from 31.7% and 41.0% of the control stump and root samples, respectively; the figures for Basamid treatment were 5.8% and 11.5% and for Sistan 0.8% and 1.8%, respectively. After 3 years, the values for the stump samples were: control 60.8%, Basamid 11.6% and Sistan 2.5%, and those for the root samples were: 39.7%, 16.5% and 0%, respectively (Figure 23.3). Experiment 2 on Scots pine gave similar results, Sistan again being appreciably better than Basamid (Figure 23.4). In Experiment 3 on Scots pine, methyl bromide almost completely eliminated *Armillaria* (Figure 23.5); Sistan gave good control in the stumps, but, as judged from the 3-year sample, was less effective in the roots.

The control stumps and roots in these three experiments were, in general, extensively decayed by *Armillaria*, especially in the sapwood. Occasionally, rhizomorphs were present in the bore holes. The characteristic creamy-white mycelium of *Armillaria* was common under the bark on the stump and most roots. In contrast, the wood of the Sistan-treated stumps was usually quite sound, with little or no decay. Although rhizomorphs were quite common beneath the bark, sub-cortical mycelium was conspicuously absent.

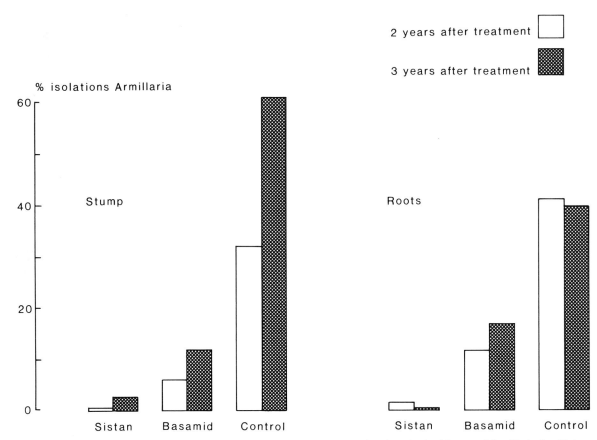

Figure 23.3 Results from Experiment 1 on Lawson cypress stumps showing the incidence of Armillaria for Sistan, Basamid and control treatments.

There was evidence for a marked increase in the incidence of *Armillaria* in the body of the control stumps between the 2- and 3-year samples, but little evidence of a change in the roots (Table 23.3).

Trichoderma spp. were commonly isolated from the stumps treated with Sistan and Basamid, in contrast to the control stumps (Figure 23.6). Thus in Experiment 1, 51.7% of the isolations from Sistan-treated stumps and 16.6% of those from Basamid-treated stumps yielded *Trichoderma*, compared with only 2.5% for the control stumps.

Experiments 4, 5 and 6 (Table 23.1) gave similar results and are not described in detail here.

In Experiment 7, Sistan significantly reduced the incidence of *Armillaria* in all five species in both the stump and root samples (Table 23.4).

Table 23.5 summarises the data from all seven experiments for *Armillaria* and *Trichoderma*.

Table 23.3 Incidence of *Armillaria* (%) isolated from control stumps

Time after fumigation:	Stumps		Roots	
	2 years	3 years	2 years	3 years
Experiment number*				
1	31.7	60.8	41.0	39.7
2	23.3	30.8	38.2	22.3
3	19.2	50.8	30.2	48.1

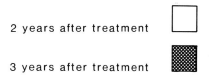

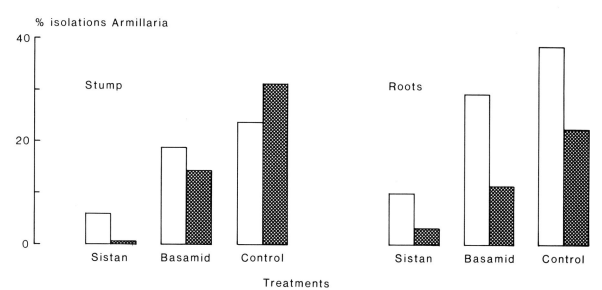

Figure 23.4 Results from Experiment 2 on Scots pine stumps showing the incidence of Armillaria for Sistan, Basamid and control treatments.

In total, 95 stumps were treated with Sistan and from the stump body only 2% of the isolations yielded *Armillaria*, compared with 36.7% isolations for the same number of control stumps. The equivalent figures for stump roots were 10.6% and 45.3%. *Trichoderma* was isolated from just over half the Sistan samples but only 20% of the controls. The results for Basamid were considerably less good although they were still significantly better than the controls.

Discussion

The results show that fumigants have considerable potential for reducing the *Armillaria* inoculum in the ground in situations where it cannot be physically removed. Of the three treatments tested, methyl bromide gave the best results, almost completely eliminating *Armillaria* from the treated stumps. However, the highly toxic nature of the product prevents its use except by licensed operators and thus greatly reduces the

Table 23.4 Incidence of *Armillaria* (%) in Sistan-treated and control stumps and roots in five conifer species. Experiment 7.

Species	Isolations with Armillaria (%)			
	Stump		Roots	
	Sistan	Control	Sistan	Control
Douglas fir	4.5	40.0	8.4	53.0
Grand fir	1.5	32.0	1.9	34.2
Japanese larch	0	36.0	9.2	44.4
Serbian spruce	5.0	36.5	11.1	36.3
Western hemlock	1.5	37.5	5.0	40.7

Table 23.5 Summary of results from Experiments 1–7 on fumigation of tree stumps.

Treatments	Number of stumps	% Armillaria		% Trichoderma	
		Stump	Root	Stump	Root
Sistan	95	2.0	10.6	54.4	55.4
Basamid	36	10.1	19.3	27.6	42.0
Control	95	36.7	45.3	20.2	19.8

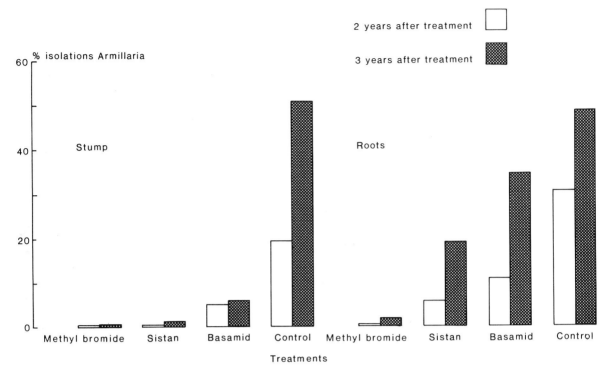

Figure 23.5 *Results from Experiment 3 on Scots pine stumps showing the incidence of* Armillaria *for methyl bromide, Sistan, Basamid and control treatments.*

scope for its use on amenity trees. Sistan gave consistently good results in the trials, being especially effective in reducing or eliminating *Armillaria* from the stump body, but also significantly reducing the incidence in the roots. Basamid appeared to be restricted in its diffusion through the stump; it being frequently noted that zone lines had developed around the bore holes, indicating that only partial control was occurring. The problems with Basamid are probably related to the fact that it is a powder rather than a liquid, requiring moisture for activation. An additional trial has been set up to test Basamid applied to stumps as a water-based paste.

The role played by *Trichoderma* species in eliminating fungal pathogens from fumigated wood has long been a subject of interest (Bliss, 1951). There was a high incidence of *Trichoderma* spp. in the fumigated stumps (Figure 23.6), but from the work conducted here it was not possible to determine how far an antagonistic effect of *Trichoderma* was supplementing the direct fungicidal agents of the fumigation.

In the USA studies (Filip and Roth, 1977), the fumigated stumps were excavated one year after treatment. In our trials, the stumps have been left *in situ* for up to 3½ years after treatment and, in Experiments 5 and 6, stumps will continue to be sampled at intervals for up to 5 years. Although the incidence of *Armillaria* in the control stumps has increased with time (Table 23.3), the effect of the fumigants has not diminished. Even 3½ years after treatment many of the Sistan-treated stumps appeared to be in an almost pristine condition. Although rhizomorphs frequently grow under the bark on these stumps, mycelium is rarely present and the wood is apparently not penetrated by the fungus. The evidence, so far, suggests that the stumps will remain free from *Armillaria* for several years, but the long-term effect is unknown.

The overall incidence of *Armillaria* is probably higher than indicated by the data from cultures as not all isolations attempted from

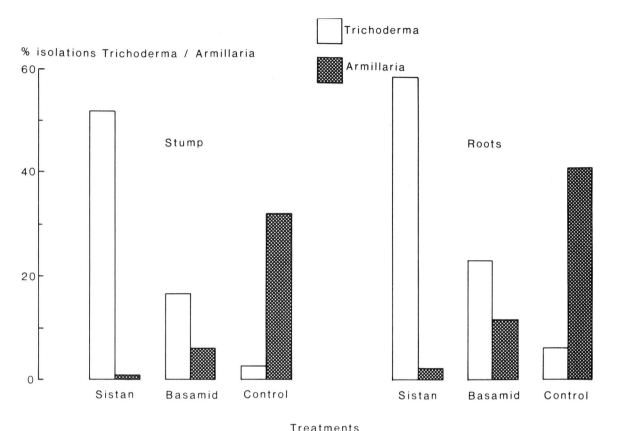

Figure 23.6 *Results from Experiment 1 showing the incidence of* Trichoderma *and* Armillaria *for Sistan, Basamid and control treatments.*

decayed wood actually yielded *Armillaria*. Although it is not possible to quantify this under estimation, it is likely to be greater for the control stumps, where frequently the decay was in an advanced state and bacteria and other contaminants may have masked the *Armillaria*. Only two other decay organisms were isolated, *Hypholoma fasciculare* from several stumps of Lawson cypress and Scots pine and *Heterobasidion annosum* from two stumps of Serbian spruce. Bacteria, yeasts and *Penicillium* were commonly isolated from both stumps and roots.

Further trials are planned to test the fumigants on a range of common hardwood species, such as birch, oak and beech, and the existing sequential trials will be assessed in 1991.

REFERENCES

BLISS, D.E. (1951). The destruction of *Armillaria mellea* in citrus soils. *Phytopathology* **41**, 665–683.

ESLYN, W.E. and HIGHLEY, T.L. (1985). Efficacy of various fumigants in the eradication of decay fungi implanted in Douglas-fir timbers. *Phytopathology* **75**, 588–592.

FILIP, G.M. and ROTH, L.F. (1977). Stump injections with soil fumigants to eradicate *Armillariella mellea* from young-growth ponderosa pine killed by root rot. *Canadian Journal of Forest Research* **7**, 226–231.

GRAHAM, R.D. (1975). Preventing and stopping decay of Douglas-fir poles. *Holzforshung* **27**, 168–173.

GRAHAM, R.D. and CORDEN, M.E. (1980). *Controlling biological deterioration of wood with volatile chemicals*. Final Report, Electrical Power Research Institute

EL–1480, Project 272–1.

GREGORY, S.C. (1989). *Armillaria* species in northern Britain. *Plant Pathology* **38**, 93–97.

GREIG, B.J.W. and STROUTS, R. G. (1983). *Honey fungus*. Arboricultural Leaflet 2. HMSO, London.

GODFREY, G.H. (1936). Control of soil fungi by soil fumigation with chloropicrin. *Phytopathology* **25**, 246–256.

HOUSTON, D.R. and ENO, H.G. (1969). *Use of soil fumigants to control spread of* Fomes annosus. Forest Service Research Paper NE–123. United States Department of Agriculture.

KUHLMAN, E. G. (1966). Recovery of *Fomes annosus* spores from soil. *Phytopathology* **56**, 885.

RACKHAM, R.L., WILBUR, W.D., SZUSKIEWICZ, T.E., and HARA, J. (1968). Soil desiccation and fumigation for *Armillaria* root rot in citrus. *California Agriculture* **22**, 16–18.

THIES, W.G. and NELSON, E.E. (1982). Control of *Phellinus weirii* in Douglas-fir stumps by the fumigants chloropicrin, allyl alcohol, Vapam or Vorlex. *Canadian Journal of Forest Research* **12**, 528–532.

Arboriculture safety

Paper 24
Safety of harness and sit-belts in tree surgery work

H. Crawford, Principal Scientific Officer, National Engineering Laboratory, Department of Trade and Industry, East Kilbride, Glasgow, G75 0QU, U.K.

Summary

The safety aspects of an investigation of harnesses and techniques for tree climbing are summarised and the merits and disadvantages of sit-belts and harnesses are discussed in relation to working comfort and safety. A range of drop tests on sit-belts and harnesses, using articulated dummies, is described, and results of fall-arrest force trials are analysed. The influence of future European and International standardisation work is considered.

Introduction

The National Engineering Laboratory (NEL) was asked by the Health and Safety Executive to investigate harnesses and techniques for tree climbing, with particular emphasis on safety aspects and the 'working day' merits and disadvantages of harnesses and 'sit-belts'. The object was to identify and study state-of-the-art equipment available to arboricultural contractors, local authority direct works departments, self-employed tree surgeons and the Forestry Commission.

NEL staff visited several colleges which run training courses for arboriculturists and gained a broad picture of recommended equipment. Forestry and arboricultural applications used by the Forestry Commission were also studied. The strict policy of the UK Forestry Commission is that all tree climbers must wear full harness with compatible climbing aids and equipment, but most arboriculturists prefer to use a sit-belt.

Advocates of sit-belts consider that the full harness is cumbersome, restrictive, uncomfortable and more likely than a sit-belt to 'snag' branches. Wearers of full harness commented on the greater safety of the harness during a fall. The accident statistics do not confirm either argument. UK and international accident reports indicated that the profession of tree climbing is not accident-ridden. In common with other industrial situations, the accidents invariably happen to people who wear neither belt nor harness, or wearers of such equipment who, at the time of the fall, are not connected to an anchorage, i.e. a tree.

With such a paucity of field data, it was necessary to carry out much of the investigation in the laboratory. This paper reports some of the findings of that laboratory work.

Basic observations prior to laboratory study

Several visits to colleges, arboreta and forest workings provided a broad view of the various operations performed by tree surgeons, cable crane loggers and seed collectors. From observations it is clear that many tasks undertaken by arboriculturists call for more movement in trees in a working day than would be required of cable crane loggers.

Considering the range of tasks undertaken by arboriculturists, tree surgeons' usual working preference for body support of the sit-belt type is understandable. When climbing a tree with a climbing rope, the action of the worker is to 'thrutch' upwards. This term describes the movement of pulling with both arms on the 'down' side of the rope while heaving upwards with the hips and legs and adjusting the prussik knot upward. This rhythmic climbing action

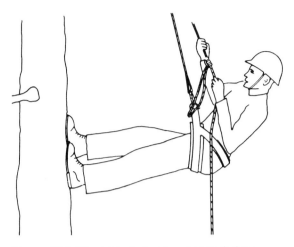

Figure 24.1 *'Thrutching' climbing rope technique.*

can only be pursued easily if good foot contact is maintained with the tree. This can only happen if the point of suspension is somewhere close to the centre of gravity (CG) of the operator. For most people, the CG in the vertical plane lies between the navel and the crotch, in the region of the pelvis; thus, the most comfortable thrutching suspension system is one that has the line of force close to the pelvis as shown in Figure 24.1.

Similar reasoning can be applied when a tree surgeon is suspended by a climbing rope from a high point in the tree and is required to move out on a branch. The closer the suspension point to the CG of the surgeon, the lower the component of body weight applied to the branch by the feet. The argument can also be applied to tree climbing with strops. Here, the mode of climbing often requires the climber to lean backwards into the body support in a fashion resembling the working position of an electricity industry linesman. The linesman's solution is the time-honoured polebelt with polestrap, not very different in concept from a tree climber's sit-belt.

Thus, a study of work tasks and procedures can justify the present preference of the many arboriculturists who use sit-belts. Only recently have full harnesses designed specifically for tree climbers been produced. These, of course, contain more component straps than sit-belts, an objection raised by many of those interviewed, but they do allow tree surgeons the required working position and are inherently safer in the event of a fall.

Tree-climbing staff in the Forestry Commission agreed with the Commission's policy directing that full harness must be worn. The procedures and techniques for seed collectors have been published and practised since at least 1965 (Seal *et al.*, 1965) and include the use of full harness.

Observations at several colleges and sites indicated that tree-climbing workers are mostly within a narrow range of body sizes, and possess a fairly high degree of physical fitness. The body sizes were clearly well within the 0.5–99.5 percentile body sizes listed in the best known anthropomorphic studies.

Table 24.1 Body dimensions (mm) of six men in laboratory study of sit-belts and safety harness.

	Subject number					
	1	2	3	4	5	6
Overall height in boots	1600	1715	1765	1830	1905	1945
Height to crotch	765	790	850	850	940	915
Chest girth	865	1070	990	990	940	1130
Waist girth	765	965	865	940	865	1015
Buttock girth	865	1040	990	1055	990	1070
Position of centre of gravity from boot soles	965	1005	950	980	1115	1155
Weight in clothes (kg)	50.5	82.3	73.6	90	70.9	105
(lb)	111	181	162	198	156	231
Thigh girth	470	585	545	615	520	645
Seated height	865	930	900	965	950	1000
Arm length	700	700	745	790	865	850

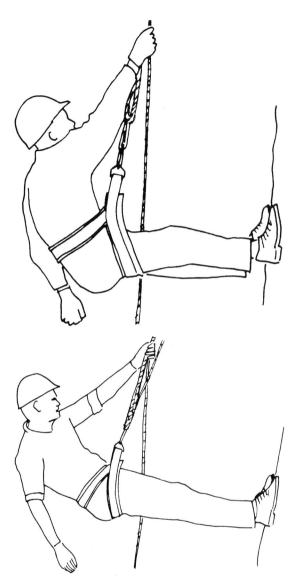

Figure 24.2 Sit-belts for tree surgeons, showing (a) small man at 90° and (b) arrangement of straps on medium to large men.

Laboratory study

In the laboratory, we studied six men. The largest was at least as large as the men seen on site, but the smallest was smaller than any seen in the field (Table 24.1). Four 'makes' of sit-belt and five of full harness were used.

Sit–belt comfort

All six men agreed that the fitting instructions and procedures for the four belts were ex-

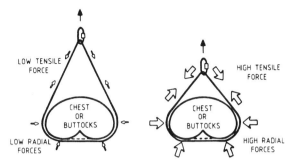

Figure 24.3 Effect of angle subtended by straps joining at the 'D' ring attachment on strap and radial forces in sit-belts for tree surgeons.

tremely simple to apply and that the freedom of movement and general manoeuvrability, whilst sitting suspended, was good for all four models. The men agreed that the position of the 'D' ring attachment for all four sit-belts ensured comfortable reach when climbing with the rope and prussik knot technique. Some of the sit-belts tended to suspend the wearer from somewhat below his CG. In the course of a working day, this poses no problem because the wearer simply sits up slightly (no conscious effort required) to attain a shift of CG and consequent equilibrium.

Views on day-long comfort were much more varied. The smallest subject (No. 1) had difficulty with the geometry of all four sit belts. A common feature of all models tested was that the waist strap was jointed to the sit-strap at an angle of 90° (see Figure 24.2a), but the overall size and distribution of straps was arranged more for middle–large-sized men (compare Figure 24.2b). Hence, subject No. 1 found that either the 'waist' strap was too high and caused discomfort at the lower ribs, or the sit strap was too low and led to 'clamping' of the thighs.

Two of the sit-belts had another design problem. The comments of subjects 2–6 confirmed that the angle subtended by the straps joining at the 'D' ring attachments was important in determining comfort. Several of the subjects complained of discomfort and pain at the waist, hip bones and thighs due to tightness or radial pressure from the straps. Figure 24.3 illustrates the effect on strap forces, and the resulting radial force, from the angle made by the attachment straps. The usual response of a designer is to in-

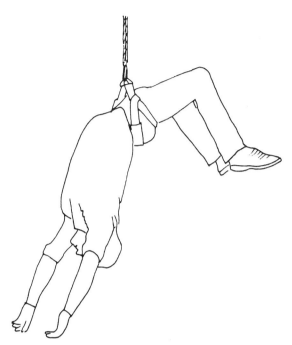

Figure 24.4 *Tree surgeon in 'jack-knife' position in sit belt.*

crease the amount of comfort padding in the painful area, but we consider that more attention should be given to means of reducing the angle.

Harness comfort

Of the five full harnesses studied and tested, two (type a) were based on successful British Standard (BS) 1397 industrial harnesses, modified for use by tree surgeons; two (type b) were from mountaineering designs but with tree surgeons in mind; and the fifth (type c) was a two-piece harness, apparently from a mountaineering design.

Some harness designs have many straps, and proper fitting of a full harness can be rather involved. It would be fair to say that a harness should be a 'one man' device.

Subjects 1 and 6, small and large men respectively, had difficulty fitting some of the harnesses to their bodies, but subjects 2–5 found that there was sufficient adjustment. Some manufacturers of harnesses offer to fit harnesses or make them to individual body sizes; this may well be a feature worth noting by prospective purchasers. Working-day comfort is highly subjective and all of the subjects commented on pressure points and working-day manoeuvrability in all the harnesses studied. The two type (b) harnesses, when properly fitted to the subjects, were considered the most comfortable and suitable for day-long working.

It was in fall-arrest safety that differences in design were more pronounced. The laboratory tests simulated suspension after a fall of each of the six subjects in the full range of belts and harnesses and a range of drop tests on each belt and harness.

Suspension trials in sit-belts

With all the sit-belts, it was obvious that the suspension point at or near the body CG would expose the wearer to the danger of rearward 'jack-knife' injuries (Figure 24.4). This is inherent in their design. Figure 24.4 also illustrates the effect of a below-CG suspension point of some of the sit-belts. The position after a fall is most likely to be head down with the spine arched backwards.

Suspension trials in harnesses

In harnesses with loose buttock straps, the straps tend to ride up the back of the wearer. In addition to the nuisance of having to reposition the strap regularly throughout a working day (a possible hazard in itself), there is the risk that, in the event of a fall, the front attachment hardware will collide with the face of the wearer. Harnesses with such buttock straps were supplied with instruction leaflets which made it clear that buttock-strap front 'D' rings must always be attached to the upper 'D' rings. We considered that a wearer could, in time, ignore that instruction and secure to the buttock strap 'D' rings only. A fall in such circumstances could lead to the position shown in Figure 24.5.

The mountaineering-type harnesses designed for tree climbers were capable of being attached to the buttock strap 'D' rings only, or to the combined upper and buttock strap 'D's. When thus combined the rope line of force for most of the six subjects tended to be somewhat higher than the CG of the wearer. In this event the wearer

Figure 24.5 *Tree surgeon in a head-first fall wearing a safety harness.*

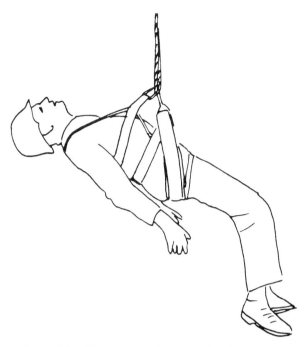

Figure 24.6 *Tree surgeon in a feet-first fall wearing a safety harness.*

feels 'feet heavy' and finds climbing arduous. Should the wearer fall while in this harness configuration, the tested harnesses showed the position after the fall to be as shown in Figure 24.6.

Many users of this type of harness are, because of body size and shape, inclined to attach to the buttock strap 'D's only, to attain a suitable climbing position. Unless there is a direct connection from shoulder strap to the attachment 'D's, the result can be a harness which gives no more fall protection than does a sit-belt.

One of the harnesses tested did have direct connection of shoulder straps to buttock strap 'D's, ensuring that, in the event of a fall, some protection was afforded against backward spinal 'jack-knifing'. The body dimensions of the largest man (No. 6) were such that the support to the upper torso via shoulder straps was reduced. Indeed, if the lower 'D's only were employed, the sudden arrest of a backward fall would lead to some 'jack-knife' arching of the spine.

Drop tests on a range of sit-belts and harnesses

All our field observations had shown that, whether climbing with a rope or strops, tree climbers work mainly with double rope, i.e. with security from two limbs of rope. All the drop tests were carried out with an articulated anthropometric dummy of 100 kg mass. A steel portal frame rig was constructed and a larch log of 250 mm diameter was secured horizontally to it, at 3.6 m centre height. The dummy was fitted with sit-belt or harness, raised to the point where its feet rested on the upper surface of the log, then the 12 mm diameter, nylon, eight-strand rope was passed from frontal attachment around the log to be secured and adjusted via prussik knot and loop to the frontal attachment once more. This simulated the normal worst-fall situation seen in the field. Each drop test was

Table 24.2 Drop tests: dummy falling backwards from log.

Test number	Harness or belt type	Strop assembly circumference (m)	Height of lower attachment from top of log (m)	Total drop distance measured after fall (m)	Peak force (kN)
1	A	2.97	0.86	2.59	6.01
2	B	2.93	1.00	2.86	6.47
3	C	2.88	1.06	2.77	5.59
4	D	3.00	0.97	2.64	6.01
5	E	2.94	0.51	2.75	6.33
6	Belt F	2.59	1.05	2.36	6.03
7	G	2.68	1.09	2.47	4.66
8	G	2.68	1.09	2.30	5.92
9	H	2.74	1.10	2.31	5.96
10	J	2.79	1.06	2.26	5.31
11	J	2.79	1.06	2.25	6.66
12	G	2.72	1.09	2.30	6.42
15	G	2.61	1.09	NA	3.72
16	H	2.64	1.10	2.30	5.55

carried out with a strain-gauged loadcell mounted at the frontal attachment and arrest force data were recorded by peak force recorder and ultraviolet recorder in tandem.

More conventional free-fall tests were conducted on a drop rig designed for testing to BS 1397; the instrumentation was similar to that in the previous tests.

Dummy falling backwards from log

On release of the dummy, we were able to replicate reliably a rearward rotational fall, the rope causing a natural restraint at near the horizontal point of the fall, and a pendulum action thereafter. Several of these tests were carried out in the presence of our physiology consultant, M. Amphoux, who vouched for the physiological accuracy of the procedure.

Except for the special circumstances of tests 7 and 15 (Table 24.2), the arrest forces were consistent within the range of 5.31–6.66 kN. Since the fall height and fall energy were virtually the same in all the tests, this narrow range in arrest force is an indication of the consistency of the prussik knot applied. However, although the prussik knot behaved as a slipping energy absorber in every case, the resulting peak force was above the medically advised limit for belts.

In the special circumstances of tests 7 and 15, we were concerned about the performance of belt G and carried out a series of confirmatory tests: tests 7, 8, 12 and 15 (see also Table 24.3, tests 13 and 14), which showed a strong tendency for this sit-belt to slide over the buttocks of the dummy and, but for the lodging of the straps in the mechanical hip or ankle joints of the dummy, the dummy would have dropped to the floor. Study of the video film of test 15 showed that the slip-out did not occur on the initial impact, but took place during a thrashing pendulum action beyond bottom dead centre of the event. Although the moulded exterior of the articulated dummy is, for durability purposes, higher in compression modulus than the buttock of man, this belt had the alarming propensity to extrude the buttocks. On a real wearer, we think the result would be just as serious. Looking at the design of this sit-belt, we noted that the sit strap exited the side 'D's at virtually 180° apart, probably the most efficient position for removing the belt when adequate downward force is applied.

Dummy falling forwards from log

The recorded arrest forces were again consistent, except for tests 14 and 22 (Table 24.3). Test

Table 24.3 Drop tests: dummy falling forwards from log.

Test number	Harness or belt type	Strop assembly circumference (m)	Height of lower attachment from top of log (m)	Total drop distance measured after fall (m)	Peak force (kN)
13	Belt G	2.70	1.09	2.23	6.93
14	G	2.61	1.09	NA	3.79
17	H	2.64	1.10	2.25	6.08
18	F	2.60	1.05	2.11	6.10
19	J	2.79	1.06	2.65	5.96
20	Harness C	2.88	1.06	2.65	5.08
21	D	2.91	0.97	2.73	7.01
22	B	2.96	1.00	2.70	3.66
23	A	2.82	0.86	2.58	5.20

14 was another slip-out of the dummy from belt G. Test 22 was on the two-piece harness and it is suspected that the prussik knot had not been sufficiently tightened prior to the drop; hence the low arrest force of 3.6 kN.

Free fall feet first, fall factor 2.0

This series of tests was carried out with procedures similar to those required for testing to British Standard 1397 (British Standards Institution, 1979). The drop was vertical from the dummy head eyebolt. The tests were carried out using 12 mm diameter, eight-strand, nylon climbing rope, arranged as it would be for a tree surgeon secured by rope and prussik, or by sling and prussik, to the branch on which the surgeon is standing, i.e. double rope at fall factor 2.0 (potential drop height divided by length of paid-out rope). This is arguably the worst fall case we can envisage and reflects a transition stage in the climbing procedure we saw at the training colleges and in the field. Instructors and operators invariably claimed that the probability of a tree surgeon 'stepping into fresh air' from this position was extremely low. We are prepared to accept that argument, but still insist that the test procedure represents the 'worst case'.

The recorded peak arrest forces were high, but not as consistent as those for the falls from the log (Table 24.4). The inconsistency appears to be due mainly to accumulated damage to the rope and prussik sling on the previous tests and to some unexplained variability in setting of the prussik from test to test.

The lowest recorded arrest force was 5.22 kN, caused by extensive slippage (0.7 m) of the prussik knot. It is coincidental that this occurred on testing belt G. Apart from this, the range was 7.95–13.13 kN and reliably reflects the degree of prussik slippage, i.e. the greater the slippage, the lower the arrest force. The results on sit-belts in tests 24, 25 and 27 were alarmingly high and cannot be justified or defended against present-day medical opinion (British Standards Institution, 1979; Amphoux, 1982; Hearon and Brinkley, 1984; Crawford, 1985, 1988).

The same comment can be made about the range of harnesses. Although most were seen to give wide load distribution through the straps, the recorded arrest force range of 7.95–13.13 kN is too high when compared with present medical opinion that 6 kN should be the arrest force limit. Thus, none of the present combinations of harness and rope/strop is adequate as a fall-arrest system and an efficient, unobtrusive, energy absorber needs to be developed.

Free fall head first, fall factor 1.0

This series of tests was carried out on the equipment described for the free-fall, feet first tests. A head-first free fall is a most unlikely event, so the drop tests were conducted at fall factor 1.0 (harness or belt anchorage adjacent to structural anchorage), the worst situation we could

Table 24.4 Drop tests: free fall, feet first (fall factor 2.0).

Test number	Harness or belt type	Strop assembly circumference (m)	Height of lower attachment from top of log (m)	Total drop distance measured after fall (m)	Peak force (kN)
24	Belt J	1.84	1.06	2.20	10.12
25	F	1.88	1.05	2.23	9.98
26	G	2.04	1.09	2.54	5.22
27	H	2.00	1.10	2.45	8.32
28	Harness C	2.16	1.06	2.40	9.15
29	E	1.68	0.51	2.51	7.95
30	B	2.22	1.00	2.60	12.16
31	C	2.16	1.06	2.49	10.31
32	E	1.68	0.51	2.36	13.13
33	A	2.18	0.86	2.42	9.66
34	D	2.11	0.97	2.45	11.28

envisage that would give rise to a head-first fall in a tree.

With a fall arrest force of 6.75–9.38 kN for the belts and 6.7–9.2 kN for the harnesses, it is clear that sit-belts cannot be defended and that the harnesses are not adequate in themselves for fall arrest on double-rope climbing procedures (Table 24.5).

Fall-arrest safety, sit-belt v. harness

The evidence and opinion of the Royal Air Force Institute of Aviation Medicine was sought for the revision of BS 1397, which warned strongly against the risks of exposing workers to arrest forces >5 kN in a waist belt and >10 kN in a full harness (British Standards Institution, 1979; Crawford, 1985).

The weight of opinion has led the International Standards Committee ISO/TC94/SC4 to recommend in the draft proposal ISO/DP 10333 (Reference N 109) that the maximum arrest force in a *full harness* should not exceed 8 kN (Crawford, 1988). Belts have been removed from the specification for Personal Equipment for Protection Against Falls, and are recommended only for use as working belts and restraint systems. A similar view now prevails within BS committee PSM/5.

Table 24.5 Drop tests: free fall, head first (fall factor 1.0).

Test number	Harness or belt type	Strop assembly circumference (m)	Total drop distance measured after fall (m)	Peak force (kN)
35	Belt J	1.84	1.30	8.64
36	F	1.88	1.50	6.75
37	G	2.04	1.30	6.93
38	H	2.00	1.19	9.38
39	Harness C	2.16	1.45	8.55
40	D	2.11	1.48	8.64
41	B	2.22	1.56	6.79
42	A	2.18	1.49	9.20
43	E	1.68	1.97	6.70

The same view is now reflected in the work of CEN committee CEN/TC160, where there is general accord that 6 kN should be the recommended maximum arrest force permissible in *full harness*. Here also, the subject of waist belts has been referred to the working group advising on working belts and restraint systems. CEN/TC160 is preparing the specifications mandated for introduction in 1992 under the European Commission Personal Protective Equipment (PPE) Directive.

It is suggested that the established and trained working procedures of many arboriculturists expose them to falls of greater height and fall factor than could be safely endured in a sit-belt. Tree surgeons' requirement for frontal attachment compounds the problem by exposing the wearer to spinal injury and whiplash.

To recommend, however, that aroboriculturists should discard sit-belts in favour of harnesses would be presumptuous, even faulty. The widespread use of sit-belts is accepted within the profession and manifestly contributes to safer (than with no body support) working in trees. It would be reckless to condemn their use immediately.

Work is required by designers of tree surgeons' harnesses to improve their acceptability by improving comfort, reach and manoeuvrability, but above all to improve their safety features and reduce the arrest forces to within the medically recommended range. The aim should be to produce harnesses which are not just better than sit-belts, but actually do provide working suitability and proven safety. When this is achieved, it will be time to persuade the industry to convert to harnesses.

As yet, these suggestions can critically be viewed as academic, there being no substantive reports of accidents to wearers of belts or harnesses. This must be acknowledged, but does not detract from the wealth of medical advice to employ full harness where exposed to risk of substantial fall or fall factor. It will be interesting to see whether the introduction of the PPE Directives in 1992 will force this change upon the profession.

ACKNOWLEDGEMENTS

Appreciation is expressed to the Health and Safety Executive for their permission to quote from the much fuller work (Crawford *et al.*, 1990).

This paper is presented by permission of the Director, National Engineering Laboratory, Department of Industry. It is British Crown Copyright.

REFERENCES

AMPHOUX, M. (1982). Physiological constraints and design of individual fall protection equipment. In *Annals of the Technical Institute for Construction and Public Works*, Paper No. 401. (French, with English translation.)

BRITISH STANDARDS INSTITUTION (1979). *BS 1397 Specification for industrial safety belts, harnesses and safety lanyards*. BSI, London.

CRAWFORD, H. (1985). Fall protection in the United Kingdom – legislation, standardisation and practice. In *Proceedings of International Fall Protection Seminar*, COPE, Toronto.

CRAWFORD, H. (1988). Who's afraid of fall factor 2.0? In *Proceedings of International Fall Protection Symposium*, Orlando.

CRAWFORD, H., BRADY, A. and HARE, D. (1990). *An investigation of tree climbing harnesses and climbing techniques*. CRR 22/1990 Health and Safety Executive, Bootle.

HEARON, B. and BRINKLEY, J.W. (1984). *Fall arrest and post-fall suspension: literature review and directions for further research. AFAMRL-TR–84–201*. Air Force Aerospace Medical Research Laboratory, Wright-Patterson Air Force Base, Ohio.

SEAL, D.T., MATTHEWS, J.D. and WHEELER, R.T. (1965). *Collection of cones from standing trees*. Forestry Commission Forest Record 39. HMSO, London.

Discussion

T.H.R. Hall (Oxford University Parks).
What is the full name of the DTI document referred to and where can it be obtained?

H. Crawford.
'The Single Market – Personal Protective Equipment' published by DTI.

D.P. O'Callaghan (Lancashire College of Agriculture and Horticulture).
Currently, only Britain and France allow tree climbing. Other European countries appear to be overriding what is statistically a safe industry. Is this realistic?

H. Crawford.
Our colleagues in Switzerland and Finland insist that their tree climbers use straps and sit-belts. In the U.K., climbing is frequently performed with rope and prussik knot. As a result, safety standards have to be formed around this knowledge to ensure that tree surgery work is safe.

J. Boyd (Merrist Wood Agricultural College).
Could an energy absorber be fitted onto a safety harness without hindrance to the climber?

H. Crawford.
Yes, this can be done, but miniaturising current technology will be expensive.

A. Lyon (London Borough of Tower Hamlets).
Energy absorbers are unknown to tree surgeons. Of what do they consist?

H. Crawford.
The simplest energy absorber would be webbing sewn back to back with chevron stitching which tears when overloaded. This makes the absorber a disposable item that if 'used in anger' must be replaced with a new one.

A. Lyon.
Tree surgeons are not inactive dummies. What difference would there be in the test results if allowance was made for an alert person being able to 'protect' himself in a fall by penduluming for example?

H. Crawford.
A pendulum action lowers the fall arrest force, but equipment and techniques must consider the worst possible scenario, i.e. a taut rope system with the anchor point below the feet. As soon as the feet leave the tree, the body is in a free-fall condition before any pendulum movement commences.

R.R. Finch (Roy Finch Tree Care).
If climbing technique is modified and a taut line can always be maintained, will the arrest forces be reduced?

H. Crawford.
Yes, but it is very difficult to envisage a system where the line is always taut.

T. Walsh (Birmingham City Council).
If the 50° angle is unsuitable, why is it retained in BS 1397?

H. Crawford.
The 50° angle for industrial harnesses is somewhat historical, the view being that if it was used in the past successfully, why change without evidence to justify a change? It does not appear to be practical for tree harnesses.

M. James (Southern Tree Surgeons Ltd).
The technique commonly used was to ascend to just below, or on a level with, an anchor point and fix a strop. The rope was then thrown ahead over the next fork or anchor point. If this is practice, why assess the category 2.0 fall?

H. Crawford.
Demonstrations at the colleges showed a climber mounting the limb to which he was anchored and then throwing ahead the free end of the climbing rope.

J. Boyd (Merrist Wood Agricultural College).
This is accepted practice in training.

D.H. Thorman (David Thorman Arboricultural Services)
How can legislation for the worst case be justified in an industry with a relatively good safety record?

H. Crawford.
The brief for the work asked for a definitive answer. In the absence of statistics, some judgements are needed. The Report stresses that both belt and harness are safer than no safety, climbing or working aid.

Concluding remarks

Paper 25
A summary consideration

D.A. Burdekin, Forestry Commission Research Station, Alice Holt Lodge, Farnham, Surrey, GU10 4LH, U.K.

This conference has been about research. Although we have heard a lot about the current research programme, Professor Bradshaw emphasised the value of arboriculture to the nation, how little is spent on amenity tree research and the need to do so much more research. He reviewed various aspects of research that have been expanded by other speakers, but one particular aspect has not been returned to. There are here representatives from various educational establishments, and they should note that there is a need for good arboricultural courses and more of them to meet the needs of the industry.

The first session of papers was about the storms of 1987 and the storms earlier this year. The Forestry Commission speaker, Chris Quine, talked about return time of the storms experienced. The fact that we had three gales in 2 years did not mean that this was likely to repeat itself, a longer-term view had to be taken. There was a discussion about broadleaves and whether the 1987 gale was damaging because trees were in leaf at the time.

John Gibbs gave a very eloquent exposition on the surveys which he and his colleagues had conducted in parklands to record the species of tree which were blown down and the parts of the tree which were affected, and whether or not decay was present. Hilary Bell explained the objectives of her study and showed some superb photographs of windblown trees. David Cutler reported on the Task Force Trees and R.B.G. Kew roots survey and concluded by proposing a couple of research programmes. One project he called rescue archaeology, that is, being opportunistic and making sure all the information possible was collected from windblown trees – I go along with that, as I was involved in similar work following the 1976 windblow. He also emphasised the importance of managing all trees and especially those that have been maltreated. It was clear during the surveys carried out that it is very difficult to determine whether an observation relates to cause and effect or whether it is merely a coincidence. It is time that we created experimentally some of those situations where trees blow over, for example near to ditches. Also, the effect of planting trees alongside severely compacted edges, such as roadsides, might be assessed by pulling trees over.

Soil compaction experiments and consideration of the effects of waterlogging were reported by Ben Hunt. The need for making larger planting pits for urban sites was emphasised as was the importance of really good ground preparation prior to tree planting. It was interesting to hear Steve Colderick from the Forestry Commission talking about his urban tree study. He emphasised the significance of damage to trees by mowers and strimmers, I have seen the amount of damage done by careless use of these and other machines. Tree nutrition was mentioned as being important, but the debate remains as to whether nutrition or soil aeration is most important. We really ought to know more about the sites into which trees are to be planted.

It is very easy to say you plant into a pavement – what is the soil underneath that pavement? What was the previous treatment in the formation of that road or little triangle? What is its status in terms of nutrients and drainage? There may be similarities between urban planting sites and reclaimed land where soil has been moved on a vast scale. Giant machines moving around a site prior to planting will totally destroy natural drainage. There have also been observations on the low nutritional status of the soil on reclamation sites. Poor soil aeration and low nutritional status may be suspected but we do not assess the site before we plant in it. Wouldn't it be valuable to insert some iron rods into the soil 6 or 9 months before the trees are to be planted, to assess the prevailing conditions? Prescriptions could then be formulated with confidence and the investment will be well made.

John Good and John Innes gave valuable reviews of the present position on air pollution and tree health. The 'green movement' is actively pressing politicians for reductions in pollution. Those of us in tree research have been very careful to ensure that there is genuine and good evidence for or against the role of pollution. This has brought politicians and researchers together and increased awareness about trees. Steve Hull's paper on ash dieback reflected my earlier comments about the effect of soil compaction, chemicals and cultivation on tree roots in the rural scene.

Successful tree establishment depends on tree quality, as described by Roger Bentley. Semi-mature trees continue to be used, and Colin Norton worked opportunistically at the Glasgow Garden Festival. Simon Hodge reported that he had detected no effect on tree establishment from the use of water-retentive materials he tested. Tim Walmsley's research with different water-retentive polymers contrasts with Simon Hodge's results – the water-retentive materials that Tim tested did give benefits. As in this case, researchers do not always agree. An ongoing debate on water-retentive materials should expose the arguments for and against their use. However, it does make it difficult for you as practitioners to go away and decide whether to use these materials or not. If I were in your shoes, I would be very tempted to try them on a proportion of trees in a scheme and to see whether you think they work or not. There may be circumstances in which these materials work and some in which they do not.

The paper by John Handley about amenity tree management was fascinating. He emphasised the sorts of thing we must all do; we must have some sort of database of our resource, otherwise how on earth can we manage it? Somehow, one has to find time for creating the database.

David Seaby's review of decay detection provided an invaluable resumé of techniques. Wound treatments have been part of the research programme since 1976. Both Peter Mercer and David Lonsdale have had a lot of influence on tree surgery practice. Martin Dobson's presentation gave a very clear picture of salt damage to trees with some interesting results and practical implications. John Turner reviewed watermark disease and warned about planting too much *Salix alba*.

Robert Strouts reviewed a number of disease diagnoses that he had been engaged in, to illustrate the difficulties in the necessary detective work. Russell Matthews questioned why people plant trees and what is the value of the tree, and I do look forward to hearing his conclusions at the end of the study. Brian Greig has been involved with honey fungus for a long time and it is interesting to see how the work has developed. Now there appears hope that stump treatment may eventually be possible.

Practitioners tend to be impatient for practical results, but research does take time. However, I find myself, like the practitioners amongst you, pressing researchers for answers; I really want to know what they think is going to be the result of their experiments. I know you feel slightly exasperated that you are not getting blueprints for planting, maintenance, protection, or whatever it is, as quickly as you would like them. However, I am sure that, as during this seminar, they will give you – at a moment in the course of their research – their best estimate of what you should do. Your interaction with researchers is very important and welcomed. Meetings such as these provide invaluable opportunities to interchange ideas, results and enthusiasms so that researchers and practitioners are not working in isolation.

Finally, the Minister. Clearly arboriculture does have political support. The Minister made it clear that government support for research in arboriculture is not going to be enough. Support from the industry, in whatever form that might be, is going to be essential. Industry, not only your industry, might well be able to provide some support. You may have seen in the newspapers today that one of the reports about this conference suggested a reduction in the use of peat. This could give an enormous saving for arboriculture. A number of very large commercial concerns are interested in peat. They might well play a part in financing further research. Perhaps I could make a very small appeal to the Arboricultural Association: 'the research field would welcome your support even in a small way'. The members of the Arboricultural Association and other professions that interface with trees all have a responsibility to help to finance the research needs.

Paper 26
Final comments

J.C. Peters, *Department of the Environment*

I would like to enlarge on two items. The first of these is the problem of co-ordination and liaison in the field of arboriculture, especially with so many customer groups. The Minister mentioned it, John Chaplin mentioned it and indirectly David Burdekin has mentioned the need for the whole industry to support research. There has to be a system of establishing priorities because there is so much to do and so little in the way of cash with which to do it. However, there are already some essential co-ordinating groups.

The Foresty Research Co-ordination Committee (FRCC), which is chaired by the Forestry Commission, has a Chief Scientist role within the Forestry Commission for co-ordinating customer interests in research. A year ago the Department of the Environment (DoE) set up the Arboricultural Advisory Board (AAB) of which I am Chairman. That Board has a number of members in common with the FRCC, such as the Ministry of Agriculture, the Nature Conservancy Council, the Forestry Commission, the Economic and Social Research Council and the Natural Environment Research Council. FRCC includes representation from the forestry industry, the Agriculture and Food Research Council and the universities. The AAB has representation from the Arboricultural Association, the Horticultural Trades Association, the Scottish Development Agency, the Scottish Development Department, the Welsh Office, the Landscape Institute, the British Association of Landscape Industries, the Department of Transport, the Association of County Councils and the Agricultural Development and Advisory Service.

This centralised structure gives a wonderful opportunity for the many ideas and concerns of the industry to be brought to the notice of funding agencies and researchers.

Both the FRCC and the AAB set up review groups to consider particular topics. The AAB recently set up a review group to assess the future development of the Arboricultural Advisory and Information Service.

Professor Last, whose report in 1974 recommended the establishment of the Service, is now chairing this group and is expected to report during July. I sincerely hope that this report will give an opportunity for developing the dissemination organisation to bring research results to the notice of more practitioners and to provide the feedback that the AAB needs. Without feedback, the researchers and funding agencies are working in a vacuum and can be accused of having their heads in the clouds.

The research programme that the Department of the Environment contracts the Forestry Commission to undertake was started 14 years ago. The fifth 3-year Arboriculture Research contract has been agreed with the Forestry Commission, but it is comparatively small funding, £$^{1}/_{4}$ million a year, for such a large industry. I agree entirely with Mr Burdekin's feeling that the industry itself, together with other customers, should be contributing to that core programme. The industry has the opportunity; here is a core programme with a core staff that should be built upon. I feel that the only way that we are going to convince Central Government of the genuine professional concern of the industry is through the route of professional reviews. We do not want the AAB to become just another external pressure group. So often, external pressure raises the temperature of perception of a particular concern and results in sudden shifts of funding to one area at the expense of others. So the priority must be to set the interface with policy in areas such as forestry, farm woodlands and agriculture, community forests and amenity, hedgerow trees and hedgerows – not forgetting that hedgerows contain woody plants that often have the potential to grow into trees!

Tree preservation orders were mentioned and the importance of adequate consultation with everybody with an interest in trees was emphasised. I am sure that topic will be coming forward again during the course of this summer in the form of a consultation paper. Again, there is a policy aspect and a research aspect.

My second point relates to the co-ordination of surveys. We have heard a lot about tree health surveys, ash dieback surveys and the need to try to get surveys onto some common footing which can enable comparisons to be made between them. I would like to draw your attention to a big project that the DoE and the Natural Environment Research Council are embarking upon during the course of this summer. The Institute of Terrestrial Ecology surveyed in 1978 and 1984, on a 1 km^2 basis, more than 384 squares throughout the country. By the middle of April 1990, each 1 km^2 in Great Britain will have been classified into one of 32 land use classes. This data set will make available information regarding the area in which people have a tree or a woodland or other specific interest, from the pressures that come from climate to the occurrence of roads and other features relevant to this land classification. The classification will include a lookback. In 1990, across Great Britain, surveyors will be out on the ground producing information at field level on species, types of agriculture and linear features.

The structure of this national survey will enable people at a local level to use the same methods and approach to survey their own areas of interest and then they can refer it to a national database. The results of that survey will be available by early spring 1992.

Through these two initiatives practitioners should be provided with basic information about the tree resource and have available a source of assistance to ensure that the amenity tree population is positively managed. The industry must play its role by providing support for the core initiatives funded by the DoE.

Research for Practical Arboriculture
University of York 2–4 April 1990

List of delegates

I. Adamson	English Woodlands Ltd
S. Aguss	Broadway Malyan Landscape Ltd
D. Alderman	Bedfordshire County Council
M. Allen	Norfolk College of Agriculture & Horticulture
P.J. Anderson	Broadland Tree Services
J. Angell	Department of the Environment
J. Armitage	Durham College of Agriculture & Horticulture
R.J. Ashton	Sylvan Landscapes
Mrs R. Baines	Hart District Council
G.C. Banks	Elmbridge Borough Council
I. Barnes	Four Seasons Tree Services
S. Barnett	South Glamorgan County Council
J.D. Barrell	Jeremy Barrell Treecare
E. Barrs	The National Trust
C.G. Bashford	Colin Bashford Associates
M. Bell	Mansfield District Council
C. Bennett	Milton Keynes Development Corporation
J.D. Blyth	Otley College
S. Bonvoisin	Nicholas Pearson Associates
R. Bothamley	Harrogate Borough Council
S. Bowra	Grampian Regional Council
J. Boyd	Merrist Wood College of Agriculture & Horticulture
H. Bradley	Commission for New Towns, Basildon
C. Britt	ADAS
Ms C. Brookes	Castle Point District Council
R. Brough	Northern Tree Surgeons (Scotland)
P. Brownlee	Lothian Health Board
P. Bullimore	Cambridgeshire County Council
M. Bulfin	Agriculture, Food & Development Authority
R. Byles	Custom Cutters
J.W. Callow	Gillingham Borough Council
S. Carey	Newcastle-upon-Tyne City Council
A. Carpenter	Flintshire Woodlands
N. Carter	Neil Carter Tree Surgeons
T. Casey	London Borough of Newham
M. Catchpole	Newark & Sherwood District Council
W. Cathcart	Department of the Environment
R. Chapman	Leicestershire County Council

Prof. A. Chiusoli	University of Bologna
R. Clappison	Sefton Metropolitan Borough Council
M.T. Clark	Chairman, Publicity Committee, Arboricultural Association
J. Clayton	Ryedale District Council
D.E. Coats	Boston Metropolitan Borough Council
A. Coker	Department of Transport
Miss K. Coulshed	West of Scotland College
S. Craddock	Middlesbrough Borough Council
S.K. Cresswell	Associated Arboreal Services
J. Cromar	Arboriculture Company
M. Crookes	Rochdale Metropolitan Borough Council
Ms A. Crotty	Norwich City Council
A. Das	Institute of Soil Fertility, Wageningen
D. Davies	Davies Brothers Tree Surgeons
K. Davies	Davies Brothers Tree Surgeons
C. Deacon	Self employed
T. De Keyzer	Countryside Commission
Dr M.P. Denne	University of Wales, Bangor
S. D'Este Hoare	East Hampshire District Council
P. Denton	Robert Denton Associates Ltd
Lady Janet Devitt	Devitt Garden Design
J.A. Dolwin	Dolwin & Gray
S. Dores	South Staffordshire District Council
P. Dornig	Nottinghamshire County Council
W. Downing	Nottinghamshire County Council
R. East	South Glamorgan County Council
M. Edwards	Winchester City Council
M.J. Ellison	Cheshire Woodlands Ltd
N. Elstone	Cumbria County Council
W. Engels	Department of Transport
R.C. Ennion	Kirklees Metropolitan Council
C.M. Erskine	Royal Botanic Gardens, Kew
J.M. Fahey	D.J. Tree and Landscape Specialists
M. Fauvel	Channel Island Tree Services
N. Fay	Treework Services Ltd
R.R. Finch	Roy Finch Tree Care
D. Ford	Askham Bryan College
D. Forman	London Borough of Hillingdon
A. Foster	Staffordshire County Council
P. Fountain	Northampton Borough Council
M. Fraser	Cardiff City Council
A. Free	Whipsnade Wild Animal Park
Miss E. Freeman	Rochester-upon-Medway City Council
J. Fulcher	New Forest District Council
J. Fuller	Professional Tree Services
P.W. Garner	Bournville Village Trust
S. Garside	Portsmouth City Council
R. Gaynor	Rotherham Metropolitan Borough Council
K. Gifford	Gifford Tree Service

R.D.D. Grainger	A/F Technical Services
I. Gray	North Bedfordshire Borough Council
R.M. Gray	
K.S. Green	Norwich City Council
D. Griffiths	Department of Transport
D. Gruber	Lewes District Council
E. Guillot	Prospect Tree Services
D. Hadley	Kilmarnock and Loudoun District Council
R. Hainsworth	Tewkesbury Borough Council
Miss A.M. Hall	Cumbria County Council
Dr T.H.R. Hall	Editor, *Arboricultural Journal*
F. Hanley	Oldham Metropolitan Borough Council
P. Harris	PPH Arborists
R. Harris	Cambridge City Council
J. Harrisson	Reigate & Banstead Borough Council
B. Harverson	Beeching of Ash Ltd
C.R. Hawke	Cornwall County Council
P. Haynes	Darlington Borough Council
J. Hazell	Milton Keynes Development Corporation
J.G. Hellingsworth	South Cambridgeshire District Council
M. Hemming	Coventry City Council
P. Hemsley	Askham Bryan College
S. Henchie	Royal Botanic Gardens, Kew
K. Hewitt	Eastern Landscape Service
R. Hewlett	Weymouth & Portland Borough Council
B. Higginson	Thomas Higginson & Co Ltd
D. Hill	Basingstoke & Deane Borough Council
D. Hinde	Department of Transport
M. Hinsley	East Dorset District Council
Ms S. Hodgson	Glen Kemp Hankinson
Mrs F. Holdsworth	Belfast City Council
P. Honey	Honey Brothers Ltd
R.V. Hoskins	County of Avon
S. Hunt	London Borough of Hillingdon
P. Ingham	North Norfolk District Council
C. Ingram	Merrist Wood College of Agriculture & Horticulture
Dr C. Ireland	Writtle Agricultural College
M. James	Southern Tree Surgeons Ltd
S.R.M. Jones	Richard Loader Tree Care
W.G.T. Jones	West Sussex County Council
C.Wm. Jorgensen	Arboricultural Consultant
A.D. Kendle	*Horticulture Week*
R.I. Kennedy	Midland Tree Surgeons Ltd
P. Kerley	University of Cambridge Botanic Garden
J.D. Keyes	Tree Consultancy Group
S. Killick	Newbury District Council
A.S. Kirkham	Royal Botanic Gardens, Kew
J. Kopinga	IBG 'De Dorschkamp', Wageningen
T. La Dell	Landscape Architect/Landscape Scientist

A. Laing	London Borough of Havering
M. Langley	Gwynedd County Council
Prof. F.T. Last	University of Newcastle-upon-Tyne
C. Lawrence	Southampton City Council
Miss R. Lawrence	
T. Lemon	Scottish Development Department
C. Lestrange	Bristol City Council
D. Lewis	Sheffield City Council
J.C. Lindsay	Scottish Development Agency
de G. Litchfield	Svensk TradVard 85
G. Litchfield	Treecare
D. Lofthouse	London Borough of Merton
S. Long	Epsom & Ewell Borough Council
B.G. Lousley	Berkshire County Council
V.M. Lowen	Leicestershire County Council
A. Lyon	London Borough of Tower Hamlets
J.I. Macdonald	Kirklees Metropolitan Council
A. M. Mackworth-Praed	Maidstone Borough Council
A. Mackay	Angus Mackay Landscape Consultants
I. Maclean	Scottish School of Forestry
Miss M. MacQueen	Broadland District Council
T.P. Marsh	T.P. Marsh
W.E. Matthews	Southern Tree Surgeons Ltd
J. McConville	J. M. McConville & Associates
J. McCullen	Office of Public Works, Dublin
B. McNeill	Estate & Forestry Services Ltd
R. McNeill	Estate & Forestry Services Ltd
Miss C. McPhie	Christchurch Borough Council
M. Minta	Ipswich Borough Council
I. Mobbs	Fountain Forestry Ltd
W. Moore	Arborist
P. Morgan	Royal Botanic Gardens, Kew
R. Morwood	Mid Bedfordshire District Council
B.J. Moss	Mid Glamorgan County Council
A. Motion	Rotherham Metropolitan Borough Council
A. Mouzer	East Sussex County Council
Miss C. Mullins	Blackpool Borough Council
R.M. Nicholson	Royal Borough of Kensington & Chelsea
C. Neilan	Epping Forest District Council
M. Nixon	Eastern Landscape Service
D. Noble	States of Jersey
R. O'Bryen	Devon County Council
Dr D.P. O'Callaghan	Lancashire College of Agriculture & Horticulture
M. O'Callaghan	Berkshire County Council
M. Page	Essex County Council
M. Page	Wirral Borough Council
D. Pain	Dorset County Council
R. Parker	Oakapple Landscapes
D. Paterson	Central Regional Council

N. Patrick	University of East Anglia
Miss S. Patrick	Student
R. Patton	Highland Regional Council
J.A.F. Perkins	University of Bristol
R. Perrins	Guildford Borough Council
R. Perry	R.M. Perry
K. Postlethwaite	Ashfield District Council
A.R.J. Powell	Greater Manchester Countryside Unit
J. Price	Borough Council of Wellingborough
D. Rankin	East Lothian District Council
J. Reed	Reading Borough Council
R.A. Robertshaw	Calderdale Metropolitan Borough Council
C. Robinson	London Borough of Barnet
K. Rolf	Swedish University of Agricultural Sciences
M. Roseveare	Arbor Tree Surgeons
W. Ross	Student in Rural Resource Management
L. Round	Trafford Borough Council
N. Rudd	Heriot-Watt University
D. Russell	National Trust
T.D. Russell	Forestry Commission
M. Rutherford	Neil Carter Tree Surgeons
A.P. Sangwine	Department of Transport
B. Saunders	Grampian Regional Council
R. Sim	Cumbria County Council
K. Simons	Warwickshire County Council
M. Simpson	Provincial Tree Services Ltd
Miss J. Smirfitt	Writtle Agricultural College
A. Smith	London Borough of Camden
A. Smith	Vermeer (UK) Ltd
J. Smith	Sheffield City Council
G. Souter	Norwich City Council
J. C. Spanswick	Taff-Ely Borough Council
P.E. Spurway	East Sussex County Council
F.R.W. Stevens	Arboricultural Advisory and Information Service
M.E. Summerscales	Calderdale Metropolitan Borough Council
M. Swapp	Grampian Regional Council
K. Swinscoe	Derbyshire County Council
W. Syrett	North Norfolk District Council
R. Taylor	Runnymede Borough Council
G. Thomas	Self employed
D.H. Thorman	David Thorman Arboricultural Services
S. Tompkins	Peak National Park
J. Trenchard	Honey Brothers Ltd
S.J. Verner	Lothian Regional Council
Mrs S. Wade	Purbeck District Council
D. Waldon	Royal Botanic Gardens, Wakehurst Place
R. Walker	Kingston-upon-Hull City Council
M. Waller	Kingston-upon-Hull City Council
B. Wallis	Milton Keynes Development Corporation

J. M. Walsh	Manchester City Council
T. Walsh	Birmingham City Council
P.S. Watts	Heritage Trees
D. Welham	Ipswich Borough Council
P. Wells	Barcham Trees
I. Welsh	English Woodlands
M. Welton	Scunthorpe Borough Council
J.B. Wetherell	North Bedfordshire Borough Council
J.T. White	Cheshire County Council
P. White	London Borough of Ealing
K. Wigginton	Reading Borough Council
R.S. Wiles	Luton Borough Council
C. Williams	Self employed
Miss C. Wilson	Student
S. Wingrove	Wingrove Tree Specialists
M. Woodcock	Forestry Commission
A. Worsnop	Southern Tree Surgeons (Ireland) Ltd
M. Wortley	Essex County Council
Dr C.J. Wright	Nottingham University
C. Yarrow	Chris Yarrow & Associates

Speakers

D.A. Burdekin	Forestry Commission
Miss H.J. Bell	University of Nottingham
R.A. Bentley	Pershore College of Horticulture
Prof. A.D. Bradshaw	University of Liverpool
S.M. Colderick	Forestry Commission
H. Crawford	National Engineering Laboratory
Dr D.F. Cutler	Royal Botanic Gardens, Kew
M.C. Dobson	Forestry Commission
Dr J.N. Gibbs	Forestry Commission
Dr J.E.G. Good	Institute of Terrestrial Ecology
B.J.W. Greig	Forestry Commission
Dr J.F. Handley	Groundwork Trust
D. Heathcoat-Amory, MP	Department of the Environment
S.J. Hodge	Forestry Commission
B. Hunt	University of Liverpool
S.K. Hull	Forestry Commission
Dr J.L. Innes	Forestry Commission
Dr D. Lonsdale	Forestry Commission
J.R. Matthews	Cobham Resource Consultants
Dr C.R. Norton	Heriot-Watt University
C.P. Quine	Forestry Commission
D.A. Seaby	Department of Agriculture, Northern Ireland
R.G. Strouts	Forestry Commission
Dr J.G. Turner	University of East Anglia
Dr T.J. Walmsley	University of Liverpool

Chairmen

A. Bannister	President, Landscape Institute
J. Chaplin	Chairman, Arboricultural Association
Mrs C.L.A. Davis	Department of the Environment
J.R. Fletcher	Vice-President, Institute of Chartered Foresters
S.A. Neustein	Forestry Commission
Dr R.G. Pawsey	Pathologist and Consultant
J.C. Peters	Department of the Environment
T. Preston	Chairman, Arboricultural Safety Council

Organisers

Mrs J. Berry	Arboricultural Association
Dr P.G. Biddle	Arboricultural Association
D. Patch	Forestry Commission
Mrs J. Woolford	Forestry Commission